Immunology of the Lymphatic System

Laura Santambrogio
Editor

Immunology of the Lymphatic System

Springer

Editor
Laura Santambrogio
Albert Einstein College of Medicine
Morris Park Avenue
Bronx, NY, USA

ISBN 978-1-4614-3234-0 ISBN 978-1-4614-3235-7 (eBook)
DOI 10.1007/978-1-4614-3235-7
Springer New York Heidelberg Dordrecht London

Library of Congress Control Number: 2013938836

Printed on acid-free paper

Springer is part of Springer Science+Business Media (www.springer.com)

Contents

Chapter 1
Introduction

Stanley G. Rockson and Laura Santambrogio

The disciplines of lymphatic biology and medicine are in the midst of much-needed and much-anticipated renaissance. New and expanded insights into the normal and abnormal function of the lymphatic system can be predicted to have a transformative impact upon our grasp of human physiology in health and in disease.

While it is difficult to fully comprehend the antecedents, it is unfortunately true that the lymphatic system has been the recipient of centuries of passive neglect. However, awareness of the importance of the lymphatic function to human health and disease is experiencing unprecedented growth. The lymphatic health continuum can readily be defined to encompass inflammation, autoimmune diseases, transplant rejection, cancer, cardiovascular and metabolic disorders, obesity, and many other expressions of human functional disorders.

This book delves into several concepts that have emerged in the last few years. These concepts provide a deeper understanding of lymph formation, its cellular and proteomic composition, circulation, filtration, as well as establishing a context in which to understand the defining attributes that govern the development and function of the lymphatic system in contrast to the blood circulatory system.

The last 20 years have eclipsed the notion that lymphatic capillaries are a mere footprint of the vascular system. There has been delineation of the specific growth factors and transcription factors that drive lymphatic biogenesis from endothelial cells expressing a unique lymphatic signature. Novel concepts have also emerged from the anatomy and ultrastructural morphology of the lymphatic system: these acknowledge several similarities to, but also important differences with, the blood vascular system. The concept of the lymphangion, as a functional unit defined by

S.G. Rockson
Center for Lymphatic and Venous Disorders, Stanford University School of Medicine, Stanford, CA, USA

L. Santambrogio (✉)
Albert Einstein College of Medicine, 1300 Morris Park Avenue, Bronx, NY 10461, USA
e-mail: laura.santambrogio@einstein.yu.edu

L. Santambrogio (ed.), *Immunology of the Lymphatic System*,
DOI 10.1007/978-1-4614-3235-7_1, © Springer Science+Business Media New York 2013

the lymphatic valve and characterized by the lymphatic wall capable of active spontaneous contractions, has also been further developed and elucidated. This active driving force is essential to transport lymph against the opposing net hydrostatic gradient. Important progress has been achieved in the attempts to define the cellular and matrix composition of the lymphangions and their changes under physiological or pathological conditions.

Novel insights into lymph formation have also been derived. The classical Starling principle, which states that fluid filtration rates between plasma and interstitial fluid is determined by differences in the hydrostatic and colloid osmotic pressures across the arterial and venous microvascular ends, has been revisited. Indeed, it has been shown that reabsorption at the venous end of the capillary bed is actually negligible; in fact, the greater part of tissue fluid drainage occurs through lymphatic means. This has created an even greater emphasis on the process of lymph formation and the role of plasma ultrafiltration in determining lymph composition.

A major development in the analysis of lymph composition has been made possible through the highly sophisticated and sensitive proteomic approaches developed in the last decade. Recently performed proteomic analyses have allowed a deeper insight into the "omic" composition (proteomic, metabolomic, and lipidomic) of the prenodal lymph and their relationships to those of plasma. Published analyses of lymph, under physiological and pathological conditions, reveal the rich tissue antigen composition of the lymph that supersedes the mere process of plasma ultrafiltration, reflecting the overall signature of the tissues of origin. Such analyses hold promise for the discovery of tissue-specific biomarkers under physiological and pathological conditions.

Novel insights into the cellular composition and circulation of the lymph have also been derived during the last decade. Emerging concepts regarding lymph-bound regulatory T cells and regulatory dendritic cells, and the inability to maintain immunological tolerance in mouse strains that have abrogated lymphatic transit, raise important questions regarding the role of lymph in the maintenance of immunological tolerance.

Finally, the role of lymphatic function, lymph transport, and their far-reaching implications in cancer progression and metastasis formation has begun to be elucidated.

Each chapter provides an historical prospective and a summary of current understanding of lymphatic angiogenesis, lymph formation, transport, circulation, and composition, thereby providing a comprehensive knowledge of "what is known" and "what is new" in the field of lymphatic biology.

Chapter 2
Lymphangiogenesis

Andrea M. Foskett, Sanjukta Chakraborty, and Mariappan Muthuchamy

Abstract Lymphatic vessels are intimately involved in the maintenance of tissue homeostasis, immune cell trafficking, and transport of dietary lipids. During embryonic development, growth of new lymphatic vessels or lymphangiogenesis occurs from preexisting blood vessels in a tightly regulated manner, which then undergoes remodeling and maturation to form the extensive lymphatic network. However, aberrant lymphangiogenesis is also associated with a number of pathological conditions, such as inflammatory diseases, allograft rejection, and cancer metastasis, while insufficient lymphangiogenesis underlies the debilitating condition of lymphedema. This chapter aims to provide an overview of the different cellular mechanisms and key molecular players involved in the regulation and progression of normal lymphatic vascular development (or physiological lymphangiogenesis) and pathological lymphangiogenesis. Understanding the mechanisms of lymphatic vascular development or its role in these pathological processes is a prerequisite for the efficient development of key therapeutic interventions for lymphatic-associated diseases.

2.1 Introduction

2.1.1 Structure and Function of the Lymphatic Vasculature

From a historical perspective, the first descriptions of vessels containing a colorless fluid, referred to as "white blood" or "arteries containing milk," were made as early as 300 BC (Gnepp 1984). However, it was not until 1622, that an Italian anatomist

A.M. Foskett • S. Chakraborty • M. Muthuchamy (✉)
Department of Medical Physiology, Texas A&M Health Science Center,
College of Medicine, College Station, TX 77845, USA
e-mail: marim@tamu.edu

L. Santambrogio (ed.), *Immunology of the Lymphatic System*,
DOI 10.1007/978-1-4614-3235-7_2, © Springer Science+Business Media New York 2013

and physician Gasparo Asellius observed unique vessels containing white blood in the mesentery of the dog and he called these vessels lacteals. The discovery of the lymphatic system has a longstanding history with contributions from numerous anatomists and physicians over the course of many centuries. However, since these first reports on lymphatic vessels, research on the development and function of the lymphatic system has progressed dramatically only over the last two decades, enabling a better appreciation for this complex system.

The major roles of the lymphatic system include maintenance of blood and tissue volume (Taylor et al. 1973), absorption and transport of dietary lipids from the intestine to the liver (Tso 1994), and immune cell trafficking (Angeli and Randolph 2006; Casley-Smith 1974; Johnson et al. 2006; Randolph et al. 2005). To accomplish these roles, the lymphatic system consists of a network of vessels of varying caliber that are connected through a series of lymph nodes and other lymphoid organs. The lymphatic system is responsible for the absorption of 20–50 % of the plasma volume and 50–100 % of the plasma proteins from the interstitium and drainage back into the systemic circulation daily (Yoffey and Courtice 1970). Furthermore, specialized lymphatics called lacteals present within the villi of the small intestine absorb dietary lipids that are secreted by enterocytes in the form of chylomicrons (Backhed et al. 2007; Tso 1994). In addition, lymphatic vessels function as active conduits for the passage of extravasated leukocytes and immune cells such as antigen-presenting dendritic cells, T lymphocytes, and macrophages thus representing an important step in the regulation of the immune response (Angeli and Randolph 2006; Casley-Smith 1974; Johnson et al. 2006; Randolph et al. 2005).

Lymphatic capillaries, also known as initial lymphatics, act as an entry point for interstitial fluid and macromolecules from the interstitial spaces. Anatomically, lymphatic capillaries are irregularly shaped, blind-ended, and thin-walled vessels. In order to facilitate permeability to large macromolecules and migrating cells, lymphatic capillaries are comprised of a single layer of overlapping oak leaf-shaped endothelial cells that are connected by loose discontinuous button-like junctions or flap-like valves (Baluk et al. 2007; Dejana et al. 2009). These lymphatic capillaries neither have a continuous basement membrane nor are they invested with muscle cells. To prevent the collapse of the lymphatic capillaries, they are physically tethered to the surrounding extracellular matrix (ECM) by bundles called anchoring filaments, which are composed of collagen, fibrillin, and emilin-1 (Danussi et al. 2008; Leak and Burke 1966, 1968).

The lymphatic capillaries coalesce to form a larger network of precollector vessels, leading to muscular collecting lymphatics, lymphatic trunks, and finally the lymphatic ducts (Gnepp 1984). The collecting lymphatic vessels possess a diverse structure that can differ dramatically in various tissues depending on its position in the lymphatic network. The two most distinguishing characteristics of the collecting lymphatics are the presence of numerous unidirectional bicuspid valves and varying amounts of muscle cell layers (Baluk et al. 2007; Schmid-Schonbein 1990). Lymphangions are the functional unit of the muscular collecting lymphatic that are arranged in series along the length of the vessel separated by valves (Gashev 2002). These unique structural features of the collecting lymphatics render its functional ability to transport lymph from one lymphangion to the next via active pumping mechanisms while also preventing backflow. The collecting lymphatics consist of a continuous layer of endothelial cells with “zipper-like” intercellular adherens and occludin junctions (Baluk et al. 2007; Dejana et al. 2009). A continuous basement membrane is also present in these vessels, which prevents leakage of lymph. Collagen and elastic fibers are randomly distributed in the spaces between the endothelial cells and the layers of muscle cells. Muscle cell layers are usually associated with the lymphangion segments and wrap around the endothelial cell-lined vessel wall, while usually at the valve site there are fewer muscle cells (Gnepp 1976; Gnepp and Green 1980). There exists a huge variation in the density, orientation, and organization of the muscle cells in different calibers of collecting vessels and among various species. As an example, the thoracic duct in a human has circular muscle cell layers that are oriented in a circumferential fashion (Gnepp 1984; Petrenko and Gashev 2008). However, in the rat diaphragm, muscle cells are arranged circumferentially near valve regions, while they are more longitudinally or spirally arranged between valves (Ohtani and Ohtani 2001). Remarkably, the different structural components of the lymphatic system work in concert to accomplish its principal task—the transport of lymph. The mechanisms of lymph transport have been discussed in another chapter of this book.

Dysfunction of the lymphatic system either due to genetic mutations that cause improper development or surgical procedures that damage lymphatic vessels result in a wide range of pathologies. One of the most debilitating outcomes of impaired lymph transport is lymphedema, a chronic progressive disease with no cure that is characterized by disfiguring swelling and impaired immunity. Other pathological conditions include filariasis, chylous ascites, and cyclothorax, inflammatory and autoimmune diseases, and the involvement of lymphatics as routes for tumor metastasis. This chapter focuses on the lymphangiogenesis processes in normal development and in pathological conditions.

2.2 Physiological Lymphangiogenesis

2.2.1 Embryonic Development of the Lymphatic Vasculature

American anatomist, Florence Sabin is credited with presenting the earliest and most widely accepted model of lymphatic development in 1902 (Sabin 1902, 1904, 1916). Based on elegant dye-injection experiments, she proposed that endothelial cells bud from veins to form primary lymph sacs, which in turn sprout in a centrifugal pattern to form dense lymphatic networks in surrounding tissues and organs (Sabin 1902, 1904, 1916). Several years later in 1910, Huntington and McClure argued an alternative theory suggesting that the initial lymph sacs originated from the mesenchyme, independent of the veins and only subsequently established venous connections (Huntington and McClure 1910). Over the past 100 years or so, we have moved down a rather slippery slope debating repeatedly over the origin of lymphatics. Only recently has this question been irrevocably answered with evidence from Cre/Lox-P-based lineage tracing studies by Srinivasan et al. (2007), which conclusively corroborates Sabin's model. Srinivasan et al. (2007) demonstrated that lymphatic endothelial cells (LECs) sprouted, proliferated, and migrated from venous-derived lymph sacs, giving rise to the entire lymphatic vasculature, and that hematopoietic cells did not contribute to this process (Srinivasan et al. 2007). The venous origin of LECs has also been documented in other models including *Xenopus laevis* (Ny et al. 2005) and zebrafish (Yaniv et al. 2006).

The key steps outlining the development of the lymphatic vasculature is schematically represented in Fig. 2.1. In mice, this process is initiated around embryonic day 9.0 (E9.0) when some local induction signal, albeit sill unknown, triggers the process of commitment of a few endothelial cells (ECs) lining the anterior cardinal vein, toward a unique LEC identity (Albrecht and Christofori 2011; Oliver 2004; Oliver and Alitalo 2005; Oliver and Srinivasan 2008; Tammela and Alitalo 2010). The early expression of markers such as vascular endothelial growth factor receptor-3 (VEGFR-3), also known as Fms-like tyrosine kinase 4 (Flt4) as well as lymphatic vessel endothelial hyaluronan receptor 1 (LYVE-1) identify this unique population of lymphatic endothelial fate-competent cells. However, the exact developmental, functional, or regulatory relevance of LYVE-1 is further confounded by studies in LYVE-$1^{-/-}$ mice that demonstrate the development of a normal functional network of lymph vessels and lymph nodes (Gale et al. 2007; Luong et al. 2009). These findings implicate a possible compensatory mechanism that can overcome the loss of LYVE-1.

Subsequent to the initiating step, at around E10, expression of the homeobox transcription factor prospero-related homeobox 1 (Prox1) commences in VEGFR-3^+/LYVE-1^+ cells and becomes restricted in a polarized manner to a subpopulation of ECs located on one side of the vein (Oliver 2004; Oliver and Alitalo 2005; Oliver and Srinivasan 2008). Unlike LYVE-1, the functional significance of Prox-1 in lymphatic development is well understood and remains the most valuable regulator of LEC specification and maintenance (Hong et al. 2002; Wigle et al. 2002; Wigle and Oliver 1999). $Prox1^+$ LECs bud out from the vein, migrate, and aggregate to form primary lymph sacs, eventually giving rise to a primitive lymphatic plexus. Consequently, $Prox1^{-/-}$ mouse embryos have been shown to completely lack a lymphatic vasculature, not due to an arrest in LEC budding, but rather as a result of a failure in lymphatic cell specification (Wigle et al. 2002; Wigle and Oliver 1999). On closer examination, it was apparent that the ECs that budded from the anterior cardinal vein still expressed blood vascular markers but in the absence of Prox1 failed to commit to a lymphatic identity. $Prox1^+$ budding LECs also provides feedback regulation allowing Prox1 to be continuously upregulated in the veins, which is believed to provide the spatial and temporal cues that influence the location and timing of the newly forming lymphatic primordia (Wigle et al. 2002). Also, overexpression of Prox1 was determined to be sufficient to reprogram blood endothelial cells (BECs) into LECs, by suppressing blood vascular-specific genes and up regulating LEC-specific genes (Hong et al. 2002, 2004; Petrova et al. 2002). In addition

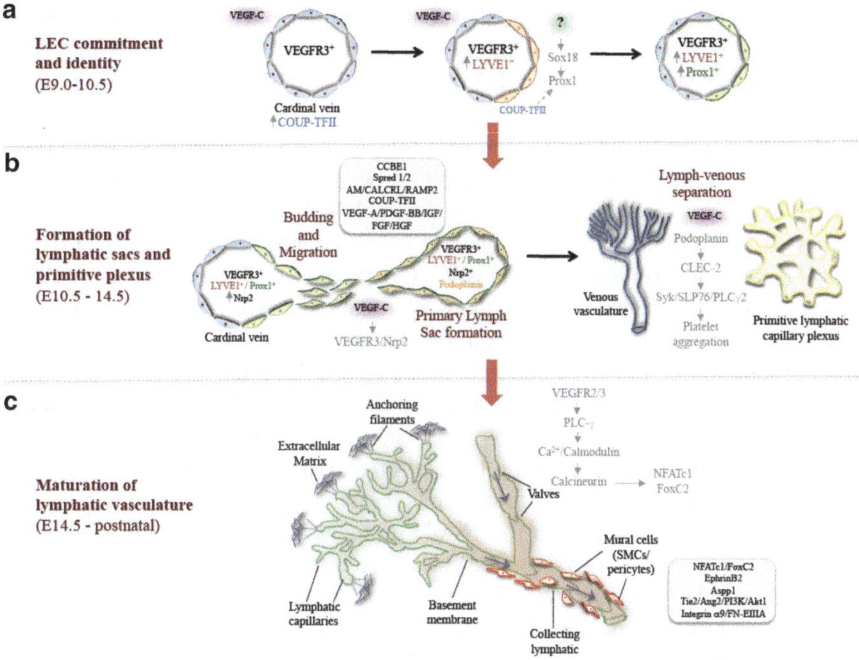

Fig. 2.1 Development of the lymphatic vasculature. (**a**) Lymphatic endothelial cell (LEC) commitment and identity (E9.0–10.5)—LECs are specified in embryonic veins, which initially express high levels of VEGFR-3. LYVE-1, the earliest known lymphatic marker, identifies a unique subpopulation of lymphatic endothelial fate-competent cells in the large veins. Following this initiation step, Sox18 induces the expression of Prox1 in VEGFR-3$^+$/LYVE-1$^+$ cells and becomes restricted in a polarized manner to a subpopulation of ECs located on one side of the vein. (**b**) Formation of lymphatic sacs and the primitive plexus (E10.5–14.5)—LEC biased cells also begin to express Nrp2. VEGFR-3/VEGF-C provides a guidance mechanism required for the budding and migration of LECs thus forming primary lymph sacs. Other lymphangiogenic factors such as CCBE1, Spred 1/2, and COUP-TFII among others have also been implicated in the sprouting potential of LECs. The LECs now begin to express podoplanin, which further activates the CLEC-2/Syk/SLP76 signaling pathway. This leads to platelet aggregation, which blocks the connections between the blood and lymphatic vasculatures, thus severing the two vasculatures from each other. (**c**) Maturation of the lymphatic vasculature (E14.5-postnatal)—Maturation of the primitive lymphatic plexus encompasses differentiation into lymphatic capillaries and collecting lymphatic vessels. The lymphatic capillaries are irregularly shaped, thin-walled vessels comprised of a single layer of overlapping oak leaf-shaped endothelial cells. The lymphatic capillaries neither have a continuous basement membrane nor are they invested with muscle cells. They are physically tethered to the surrounding extracellular matrix by bundles called anchoring filaments. The process of maturation of the collecting lymphatic vessels involves complex morphological remodeling events necessitating the occurrence of several critical steps, such as valve formation, muscle cell recruitment, and vessel specification. The calcineurin/NFATc1/FoxC2 pathway plays an important role in the formation of valves. The investiture of muscle cell layers is accomplished by several key players, such as, EphrinB2, FoxC2, Aspp1, Tie2/Ang2, and integrin α9

to regulating lymphatic specification, Prox1 is also critical for the maintenance of this lymphatic endothelial phenotype during later stages of development and in adulthood, since heterozygous Prox1$^{+/-}$ mice exhibit impaired lymphatic function and abnormalities in lymphatic network patterning (Harvey et al. 2005). These findings and those of other groups have confirmed that Prox1 is the crucial hallmark gene that is both required and sufficient to confer a LEC phenotype (Hong et al. 2002, 2004; Oliver and Detmar 2002; Petrova et al. 2002; Wigle et al. 2002; Wigle and Oliver 1999).

As a consequence, this raises the next logical question: what mechanism regulates Prox1, the master regulator of lymphatic development? The exact mechanisms upstream of Prox1 induction remain elusive, although some suggestions have been proposed in the literature. For instance, it is known that IL-3 and IL-7 can induce Prox1 expression in cultured human BECs (Al-Rawi et al. 2005; Groger et al. 2004). However, the involvement of IL-3 and IL-7 in inducing Prox1 expression in vivo is yet to be determined. As another example, Francois et al. (2008) demonstrated the direct induction of Prox1 in cultured BECs by transcription factor Sox18, while Sox18$^{-/-}$ mouse embryos lacked the development of lymphatic vasculature. Intriguingly, Sox18 expression was identified in a subset of cells on the cardinal vein believed to have a lymphatic bias, prior to Prox1 induction leading to lymphatic specification in those very same cells (Francois et al. 2008). Nuclear receptor COUP-TFII that is best known for its role in maintaining venous cell identity has recently been shown to interact with Prox1 (Lee et al. 2009; Srinivasan et al. 2010; Yamazaki et al. 2009) and was identified as a co-regulator of Prox1 function in maintaining LEC specification (Lin et al. 2010). Furthermore, the conditional ablation of COUP-TFII during early embryonic time points compromised the development of the lymphatic vasculature implicating its role in the establishment of LEC identity (Lin et al. 2010).

2.2.2 *Lymphangiogenic Growth Factors: Sprouting, Proliferation, and Migration*

Subsequent to LEC specification, the polarized budding of lymphatic fate-committed Prox1^{+} cells is driven by a receptor–ligand guidance mechanism. Although several lymphangiogenic growth factors and their receptors have been identified, the best characterized and vital to this directed migration process are vascular endothelial growth factors (VEGF)-C and VEGF-D and their cognate receptor VEGFR-3 (Achen et al. 1998; Joukov et al. 1996; Karkkainen et al. 2004; Lohela et al. 2009). Even before the onset of LEC specification by master regulator Prox1, VEGFR-3 is expressed on some ECs of the cardinal vein that are presumed to be competent to acquire a lymphatic phenotype (Albrecht and Christofori 2011; Oliver 2004; Oliver and Alitalo 2005; Oliver and Srinivasan 2008; Tammela and Alitalo 2010). During early murine development, VEGFR-3 is initially expressed by both the blood and lymphatic endothelium, but becomes mostly restricted to the lymphatic endothelium in late development and adulthood (Kaipainen et al. 1995; Wigle et al. 2002). Findings from studies using VEGFR-3$^{-/-}$ mice highlighted a role for VEGFR-3 in mediating sprouting and remodeling of the primary vascular plexus. Because VEGFR-3$^{-/-}$ mice exhibit a dramatic blood vascular phenotype and are embryonically lethal by E9.5 before the emergence of the lymphatic vasculature, its exact role in lymphatic development cannot be fully addressed (Dumont et al. 1998; Hamada et al. 2000). However, the identification of missense mutations in VEGFR-3 in patients with hereditary lymphedema (Karkkainen et al. 2000) has provided support for its role in lymphatic development.

In concert with VEGFR-3, there is now overwhelming evidence for the chemotactic role of VEGF-C as a potent inducer of lymphatic sprouting from the cardinal vein (Enholm et al. 2001; Karkkainen et al. 2004; Karpanen and Alitalo 2008; Oh et al. 1997). Findings from in vivo studies demonstrate that over expression of VEGF-C in the mouse skin results in lymphangiogenesis and hyperplasia of cutaneous lymphatics (Enholm et al. 2001; Jeltsch et al. 1997), while VEGF-C is capable of regenerating the cutaneous lymphatic network in skin of mice with lymphedema (Karkkainen et al. 2001). Furthermore, VEGF-$C^{-/-}$ embryos are perinatally lethal around E15.5 and completely lack lymphatic vessels due to defective budding and migration of the $Prox1^{+}$ LECs to form lymph sacs (Karkkainen et al. 2001, 2004). Also in zebrafish, the VEGFR-3/VEGF-C signaling axis has been implicated in the development of the lymphatic vasculature (Kuchler et al. 2006; Yaniv et al. 2006). On the other hand, VEGF-D can bind both VEGFR-2 and VEGFR-3 (Stacker et al. 1999). VEGF-D has been shown to be a potent inducer of lymphangiogenesis and is capable of mediating LEC migration (Byzova et al. 2002; Rissanen et al. 2003; Tammela et al. 2005a). In the mouse skin model, over-expression of VEGF-C or VEGF-D induced lymphangiogenesis of lymphatic capillaries, without affecting angiogenesis (Jeltsch et al. 1997; Veikkola et al. 2001). However, the lymphangiogenic potential of VEGF-D was foreshadowed by evidence in mice with a targeted inactivation of VEGF-D that developed normal lymphatic vasculature (Baldwin et al. 2005). Therefore, VEGF-D may be dispensable during the sprouting of embryonic lymphatic capillaries or perhaps VEGF-C compensates for VEGF-D in its absence.

It has been suggested that the sprouting potential of LECs is not only mediated by VEGFR-3 but also by its transmembrane co-receptor neuropilin-2 (Nrp2), which is also a receptor for class III semaphorins classically involved in neuronal axon guidance (Neufeld et al. 2002). Nrp2 selectively controls the formation of large and small caliber lymph vessels as $Nrp2^{-/-}$ mice exhibit a transient absence or severe reduction of small lymphatic capillaries during development (Yuan et al. 2002). Recent evidence further suggests that Nrp2 interacts with VEGFR-3 to promote dermal lymphatic vessel sprouting in mice in response to VEGF-C (Xu et al. 2010). Several additional growth factors have been implicated either in vitro or in vivo to be involved in various aspects of the lymphangiogenesis process namely proliferation, migration, and formation of the primitive lymphatic plexus. These include hepatocyte growth factor (HGF), fibroblast growth factor-2 (FGF-2), FGF-3, platelet-derived growth factor-BB (PDGF-BB), and insulin growth factors-1 and -2 (Auguste et al. 2003; Bjorndahl et al. 2005a; Cao et al. 2004; Kajiya et al. 2005; Kubo et al. 2002; Shin et al. 2006).

As budding and sprouting progresses to give rise to the primitive lymphatic network, the connections between the blood and lymphatic vasculatures are lost. Lymphatic capillaries begin to separate away from the veins except at the junction of the thoracic duct and the left subclavian vein. Two key players controlling this critical step are the adaptor protein SRC homology-2-domain-containing leukocyte protein (Slp76) and the tyrosine kinase Syk. Knockout of either Slp76 or Syk genes in mice resulted in abnormal blood–lymphatic connections characterized by the presence of blood-filled lymphatics and arteriovenous shunting (Abtahian et al. 2003). While closely resembling the phenotypes of Syk and Slp7 knockout mice, the genetic deletion of Spred-1 and Spred-2, members of the Spred/Sprouty family of proteins also resulted in embryonic lethality in mice between E11.5 and E15.5 that was characterized by defects in lymphovenous separation (Taniguchi et al. 2007). Likewise, varying degrees of nonseparation of the lymphatic and blood vasculatures have also been identified in mutant mice lacking several other genes such as phospholipase C gamma 2 (PLCγ2), fasting-induced adipose factor (Fiaf), forkhead transcription factor (FoxC2), and EphrinB2 (Backhed et al. 2007; Ichise et al. 2009; Makinen et al. 2005; Petrova et al. 2004). Of particular interest is the transmembrane mucin-like glycoprotein T1α/Podoplanin that is predominantly expressed in the lymphatic endothelium as early as E11 in budding LECs (Breiteneder-Geleff et al. 1999). Podoplanin$^{-/-}$ mice display defects in lymph vessel structure and function, also characterized by blood-filled lymphatics indicative of a failure in lymphovenous separation (Schacht et al. 2003). Uhrin et al. (2010) demonstrated that the aggregation of blood platelets in the connecting region between the lymph sacs and the cardinal vein is necessary for the separation of the two entities. This phenomenon that occurs at the separation zone is mediated by podoplanin upon its binding with C-type lectin-like protein (CLEC-2), which in turn is dependent on Syk, SLP-76, and PLCγ2 (Uhrin et al. 2010). These findings suggest that all these molecules work alongside each other to orchestrate a complex multistep mechanism that accomplishes lymphovenous separation.

2.2.3 Remodeling and Maturation of the Primitive Lymphatic Vasculature

Much effort has been devoted to deciphering the molecular mechanisms pivotal in orchestrating the grand finale of the lymphatic development processes that begin around E14.5 and proceed well into postnatal time points. This multistep process culminates with the remodeling and maturation of the primitive lymphatic plexus into both a hierarchically organized and a functional network of lymphatic capillaries and collecting lymph vessels (Albrecht and Christofori 2011; Makinen et al. 2007; Oliver 2004; Oliver and Alitalo 2005; Oliver and Srinivasan 2008; Schulte-Merker et al. 2011; Tammela and Alitalo 2010). Maturation of the primitive lymphatic plexus involves complex morphological remodeling of vessels necessitating the occurrence of several critical steps, such as valve formation, muscle cell recruitment, and vessel specification. One key regulator of the maturation process is FoxC2, a member of the forkhead family of proteins that is highly expressed in developing lymphatic vessels (Dagenais et al. 2004; Wijchers et al. 2006). Mutations in FoxC2 (Fang et al. 2000; Finegold et al. 2001; Traboulsi et al. 2002) were first identified in humans with the hereditary disease, lymphedema-distichiasis (LD), characterized by late onset lymphedema and venous insufficiency caused by incompetent venous valves (Mellor et al. 2007), thus sparking an interest for the role of FoxC2 in lymphatic valve formation. Similar to the pathogenesis of LD observed in humans, FoxC2$^{-/-}$ mice exhibit abnormal lymphatic vascular patterning, increased pericyte investment of lymphatics, lymphatic dysfunction, and agenesis of valves (Petrova et al. 2004). However, early lymphatic development events such as LEC specification, sprouting, and migration were not impaired in FoxC2$^{-/-}$ mice. Within the context of these findings, it is evident that FoxC2 controls only later events in lymphatic development such as valve formation and the specification of capillary versus collecting vessel phenotypes. Because mice heterozygous for both FoxC2 and VEGFR-3 were similar to the FoxC2$^{-/-}$ mice, it was suggested that an interaction exists between the two pathways and that FoxC2 cooperates with VEGFR-3 (Petrova et al. 2004). Drawing a parallel to the well-documented cardiac development process, we know that the transcription factor NFATc1 plays a pivotal role in the morphogenesis of cardiac valves (Ranger et al. 1998). It has been shown that the VEGFR-3/VEGF-C signaling axis triggers a cooperative interaction between NFATc1 and FoxC2, which in turn is essential to lymphatic valve formation and subsequently the maturation process (Norrmen et al. 2009). Another report by Bazigou et al. shed more light into the mechanism involved in the process of lymphatic valve morphogenesis (Bazigou et al. 2009). Their findings demonstrate that endothelial cell-specific deletion of encoding integrin alpha-9 (Itga9) in mouse embryos resulted in malformed rudimentary lymphatic valve leaflets that allowed retrograde lymph flow instead of the normal one-way flow of lymph (Bazigou et al. 2009). Their results further substantiated a role for the Itga9-EIIIA signaling axis in regulating fibronectin (FN) assembly, which is a requirement for the formation of valve leaflets (Bazigou et al. 2009).

EphrinB2, a transmembrane ligand of the Eph group of receptor tyrosine kinases, regulates several developmental processes including axon guidance, proliferation of neural stem cells, and angiogenesis. EphrinB2 is expressed on arterial ECs, while its receptor EphB4 is expressed by venous ECs (Adams et al. 1999; Gale et al. 2001); therefore the EphrinB2–EphB4 signaling is thought to be important for regulating arteriovenous separation within the blood vascular system (Wang et al. 1998). It is not surprising then that EphrinB2 plays an important role in the maturation of the lymphatic vascular counterpart. EphrinB2 is predominantly expressed in the LECs of collecting lymphatics and deletion of the cytoplasmic PDZ-interaction domain of ephrinB2 results in postnatal lethality in mice (Makinen et al. 2005). Mutant EphrinB2 mice exhibit hyperplasia of the collecting lymphatics, lack luminal valves in the collecting vessels, and fail to remodel their primary lymphatic capillary plexus (Makinen et al. 2005). Additionally, these mice acquire ectopic coverage of smooth muscle cells in lymphatic capillaries. These findings were similar to those observed in the FoxC2$^{-/-}$ mice discussed above, suggesting a role for EphrinB2 in the postnatal remodeling of the lymphatic vasculature that is essential to establish a hierarchically organized mature lymphatic network.

Besides FoxC2 and EphrinB2, Angiopoietin-2 (Ang-2), a ligand for the Tie1 and Tie2 endothelial-specific receptor tyrosine kinases (Saharinen et al. 2005), has also been implicated as being a regulator of the lymphatic remodeling process. Ang-2 has been shown to act as an antagonist to blood vascular stabilization, but as an agonist in lymphatic vascular stabilization (Gale et al. 2002; Maisonpierre et al. 1997). Ang-$2^{-/-}$ mice die within 2 weeks of birth and exhibit chylous ascites, subcutaneous edema, a disorganized and leaky lymphatic vasculature, impaired muscle cell recruitment to the collecting lymphatics, and defective maturation of lymphatic vessels (Dellinger et al. 2008; Gale et al. 2002). Ang-1 rescued the lymphatic defects of Ang-$2^{-/-}$ mice (Gale et al. 2002), suggesting a compensatory mechanism of Ang ligands via its associated Tie receptors (Saharinen et al. 2005; Tammela et al. 2005b). Finally, mispatterned collecting lymphatics, abnormal lymphatic vessel maturation, and subcutaneous edema associated with defective lymphatic drainage function were also reported in apoptosis-stimulating protein of p53 (Aspp1)-deficient mouse embryos (Hirashima et al. 2008). Remarkably, lymphangiography performed in adult Aspp$1^{-/-}$ mice indicated that some of the impaired lymphatic drainage was restored in spite of abnormally patterned collecting vessels (Hirashima et al. 2008).

2.2.4 Experimental Models for Physiological Lymphangiogenesis

The discovery of lymphatic-specific markers and growth factors has tremendously advanced our knowledge of the process of lymphangiogenesis using a wide array of in vitro and in vivo experimental models. LECs have been isolated from collecting vessels such as the thoracic duct and the mesentery, and from lymphatic capillaries, primarily the dermal lymphatics (Gnepp and Chandler 1985; Hayes et al. 2003; Leak and Jones 1994; Mizuno et al. 2003; Whitehurst et al. 2006; Yamaguchi et al. 2008). LEC selection and separation from BECs is now performed using fluorescence-activated cell sorting (FACS) or magnetic bead approaches (Hirakawa et al. 2003; Kriehuber et al. 2001; Podgrabinska et al. 2002). Podoplanin, LYVE-1, and CD 31 are widely used as selection markers in these techniques. However, while employing these isolation techniques in lymphangiogenesis models, the inherent heterogeneity in morphological and functional characteristics of LECs grown from initial versus collecting lymphatics must not be overlooked (Kawai et al. 2008).

In vitro models of 2D and 3D LEC cultures allows the study of early mechanisms of lymphangiogenesis, under a set of controlled and defined conditions (Bruyere and Noel 2010). Most 2D LEC cultures are seeded as monolayers on culture plates or grown on the surface of ECM-coated plates to further stimulate lymphangiogenesis. These 2D models are beneficial to tease out the individual roles of lymphangiogenic activators, but do not provide the wholesome microenvironment needed to study at large the multistep processes involved in lymphatic vessel formation.

Alternatively, 3D culture models recapitulate as closely as possible in vivo interactions between LECs and the surrounding support cells and ECM. Specifically, a 3D system enables the investigation of the early stages of lymphatic capillary formation such as sprouting, migration, and LEC morphogenic events. LECs differentiated from mouse embryonic stem cells (ESCs) can be grown as 3D spheroids called embryoid bodies (EBs) (Alajati et al. 2008). Using EB models, numerous investigators have successfully reported on the development of lymphatic-like structures when EBs were stimulated under defined conditions including growth factors, ECM, and hypoxia (Foskett et al. 2011; Kreuger et al. 2006; Liersch et al. 2006; Nilsson et al. 2004). To further mimic the in vivo context where LECs of lymphatic capillaries are exposed to shear forces induced by lymph flow through the lumen, artificial flow conditions can be superimposed to examine capillary morphogenesis (Helm et al. 2005, 2007; Ng et al. 2004). Instead of isolating LECs from a lymphatic, a novel lymphatic ring culture system employing a piece of the entire lymphatic vessel embedded in a 3D collagen matrix was designed to produce lumen-containing lymphatic outgrowths (Bruyere et al. 2008). Furthermore, Gashev et al. has also developed a long-term *ex vivo* culture system for the maintenance of lymphatic vessels (Gashev et al. 2012). This technique will enable the study of the effects of the knockdown or overexpression of genes on the functional capacity of lymph transport in isolated lymphatic vessels (Gashev et al. 2012).

In vivo models employ genetically engineered transgenic mice, particularly targeted against lymphatic-specific genes, to study different mechanisms associated with pre- and postnatal lymphangiogenesis (Carmeliet et al. 1996; Gale et al. 2007; Maisonpierre et al. 1997; Makinen et al. 2005; Petrova et al. 2004; Schacht et al. 2003; Shalaby et al. 1995; Wigle and Oliver 1999). Additionally, small animal models such as *Xenopus laevis* and zebrafish are being currently employed (Isogai et al. 2009; Ny et al. 2006; Yaniv et al. 2006).

2.3 Pathological Lymphangiogenesis

2.3.1 Lymphatic Malfunction in Primary and Secondary Lymphedema

One of the classical attributes of lymphatic vessel dysfunction manifests as primary and secondary lymphedema (Alitalo 2011; Murdaca et al. 2012; Norrmen et al. 2011; Rockson 2012; Schulte-Merker et al. 2011; Warren et al. 2007). Lymphedema is a chronic, progressive, and debilitating disease characterized by swelling in the arms and legs due to fluid accumulation in the interstitial tissue and is often accompanied by inflammation. Primary lymphedema is an inherited condition that results from genetic mutations usually present at birth but can also develop in the postnatal period. Several genes have been implicated in the development of different lymphedema syndromes (Alitalo 2011). The unifying concept in lymphedema underlies abnormalities in the lymphatic vessel development, consequently translating to the disruption in lymphatic vessel transport function. For example, early-onset congenital lymphedema such as Milroy's disease that is linked to heterozygous missense mutations in VEGFR-3 is characterized by hypoplasia or aplasia of the superficial cutaneous lymphatic capillary network (Bollinger et al. 1983; Connell et al. 2009; Karkkainen et al. 2000; Mellor et al. 2010). Another common form of primary lymphedema is a late-onset disease called lymphedema-distichiasis (LD). It is an inherited autosomal-dominant disease resulting from loss-of-function mutations in the FoxC2 gene, which is involved in the calcineurin-NFATc1 signaling pathway. The underlying pathology in human LD patients as well as FoxC2 mutant mice is defective lymphatic valve development and aberrant maturation of the collecting lymphatic vessels, thereby manifesting in clinical symptoms such as lower limb swelling and lymph reflux (Fang et al. 2000; Petrova et al. 2004). Also, a rare syndrome associated with childhood-onset lymphedema called hypotrichosis-lymphedema-telangiectasia is caused by mutations in the Sox18 gene that is upstream of Prox1 (Irrthum et al. 2003). Yet another form of lymphedema called Hennekam lymphangiectasia-lymphedema syndrome has been associated with mutations in collagen and calcium-binding EGF-domain-1 (CCBE-1), which has been reported to be essential to LEC sprouting (Alders et al. 2009). Recently, connexin 37 and 43, downstream of FoxC2 has also been implicated in the pathology of lymphatic disorders such as lymphedema and cyclothorax (Kanady et al. 2011). Other novel mutations identified in the pathology of hereditary lymphedema include the gap junction protein GJC2 that encoded connexin 47 and protein tyrosine phosphatase PTPN14 (Au et al. 2010; Ferrell et al. 2010).

Secondary lymphedema on the other hand is acquired and results in lymphatic vessel damage and dysfunction as a consequence of surgery, trauma, and infections such as filariasis or inflammation, or radiation therapy. Lymphatic filariasis also known as elephantiasis is a common form of secondary lymphedema. In this disease, infection of the lymphatic vessels is caused by mosquito-borne parasitic nematodes *Wuchereria bancrofti*, *Brugia malayi*, or *Brugia timori*. During chronic filarial infection, toxins released by the dead or adult worm triggers a massive inflammatory response, causing an increased production of proinflammatory cytokines and immune cells, which act directly on the lymphatics and are believed to contribute to a sequential change in its architecture. The lymphatic vessels harboring the various stages of the filarial parasite gradually become dilated with nonfunctional valves, impaired contractility, and abnormal drainage patterns. As a result there is massive fluid accumulation with resultant lymphedema and chronic obstructive lesions in the lymphatic vessel wall that cause severe and irreversible damage to the function of the lymphatics (Bennuru and Nutman 2009; Dreyer et al. 2000). Furthermore, recent evidence suggests that these filarial parasites are capable of inducing LEC proliferation and lymphatic remodeling (Bennuru and Nutman 2009; Pfarr et al. 2009). Another leading cause for secondary lymphedema is following radical axillary lymph node dissection during breast cancer surgery. Following this surgical procedure, the removal of lymph nodes combined with radiotherapy effectively destroys the lymphatic vessel network thus impairing lymph transport especially through the collecting lymphatics (Murdaca et al. 2012; Stanton et al. 2009).

Currently, there is no absolute cure for lymphedema. However, there are limited treatment options such as physiotherapy, lymphatic massage, and compression bandages that offer mediocre management of the symptoms, particularly in the reduction of swelling. Of interest, in recent decades, an explosion in the unraveling of various mechanisms that regulate the process of lymphatic vessel development has provided the basis for lymphangiogenic factors as a treatment modality for both primary and secondary lymphedema. Several different animal models have been employed to study lymphangiogenic therapy (Alitalo 2011; Norrmen et al. 2011). As one of many examples, VEGFR-3 therapy, via the delivery of excess of VEGF-C, was employed in a mouse model of Milroy's disease to rescue defective lymphangiogenesis (Karkkainen et al. 2001). VEGF-C has been administered as a therapy in many different ways such as recombinant proteins, adenoviruses, and adeno-associated viruses (Saaristo et al. 2004; Szuba et al. 2002; Yoon et al. 2003). Results from many laboratories indicate that VEGF-C is a promising therapy to promote lymphatic capillary formation and reduce edema to some extent (Saaristo et al. 2004; Szuba et al. 2002; Yoon et al. 2003). Furthermore, collecting lymphatic vessels which are mostly damaged in secondary lymphedema were investigated in a mouse model of axillary lymphadenectomy following axillary lymph node dissection and removal of the associated collecting vessels (Tammela et al. 2007). This study demonstrated that the combination of VEGF-C and VEGF-D treatment with lymph node transplantation promoted the robust growth of lymphatic capillaries that eventually remodeled into collecting lymphatics and even fused with the transplanted lymph nodes thus reducing edema (Tammela et al. 2007). In addition, VEGF-C and VEGF-D therapy has been effectively employed in large animal models such as pigs to increase lymphatic function and therefore reduce edema via the growth of new lymphatic vessels (Lahteenvuo et al. 2011). Furthermore, VEGF-C and Ang-2 was delivered into sheep using a gel-based drug delivery system after the removal of a single popliteal lymph node (Baker et al. 2010). The findings of this study demonstrated significantly reduced edema associated with an improvement in lymphatic function (Baker et al. 2010).

2.3.2 Lymphangiogenesis in Inflammation and Immune Dysfunction

Increased lymphangiogenesis and remodeling of lymphatic vessel networks are a well-recognized outcome in both acute and chronic inflammation (Cueni and Detmar 2008; Jurisic and Detmar 2009). During inflammation, lymphangiogenesis plays a pivotal role in facilitating the resolution of edema and the mobilization of leukocytes and immune cells such as macrophages and dendritic cells. Notably, pro-inflammatory signals lead to the production of VEGF-A, VEGF-C, VEGFR-3, Prox1, and NF-κB by a variety of cells, indicating a role for these factors in lymphatic vessel formation during the inflammatory process (Flister et al. 2010; Mouta and Heroult 2003; Ristimaki et al. 1998). In a mouse model of peritonitis, inflammation has been shown to induce lymphangiogenesis via the upregulation of VEGFR-3 (Flister et al. 2010). During acute inflammation, macrophages are recruited to the site of injury (Kang et al. 2009; Kerjaschki 2005). These recruited macrophages are capable of transforming from naïve monocytes into VEGF-C/D producing cells thus stimulating lymphangiogenesis (Schoppmann et al. 2002). Macrophages have also been shown to directly transdifferentiate into LECs (Maruyama et al. 2005). Furthermore, the activation of the NF-κB pathway in LECs upregulates Prox1 and VEGFR-3, which renders the lymphatic vessels more sensitive to VEGF-C and VEGF-D produced by leukocytes (Flister et al. 2010).

There is still an open debate about whether lymphangiogenesis promotes or hinders the resolution of edema and inflammation. One school of thought is that increased lymphangiogenesis might have beneficial effects on resolving inflammation by the clearance of edema, inflammatory cells, and cytokines and that it typically occurs at sites of tissue inflammation, for example, in immunization and in bacterial infection (Alitalo et al. 2005; Baluk et al. 2005). Remarkably, in a mouse model of chronic airway inflammation induced by *Mycoplasma pulmonis* infection and subsequent TNF-α production, VEGF-C/D producing inflammatory cells drove lymphangiogenesis thus promoting fluid clearance and the resolution of inflammation (Baluk et al. 2005). When the lymphangiogenic process was blocked with antibodies to VEGFR-3, the severity of the inflammatory response was exacerbated leading to bronchial lymphedema (Baluk et al. 2005). Interestingly, Okazaki et al. (2009) demonstrated that suppressing integrin $\alpha 5\beta 1$ signaling with small-molecule inhibitors inhibits lymphangiogenesis in the same model presumably by direct inhibition of migration and proliferation of LECs. Additionally, it is believed that in animal models of chronic arthritis and in human rheumatoid arthritis, the formation of new lymphatic vessels prevents the accumulation of inflammatory cells thereby improving swelling of the inflamed synovial joints (Polzer et al. 2008).

However, in mouse models of UVB-irradiation-induced chronic skin inflammation resembling psoriasis in humans, Kajiya et al. demonstrated increased lymphatic hyperplasia associated with VEGF-A (Kajiya et al. 2006). Further analyses using this model showed that the blockade of VEGFR-3 prolonged UVB-induced chronic skin inflammation and that lymphatic hyperplasia was associated with downregulation of VEGF-C, which was accompanied by infiltration of macrophages (Kajiya and Detmar 2006; Kajiya et al. 2009). Consequently, intradermal injection of VEGF-C attenuated UVB-induced inflammation and edema formation by promoting lymphangiogenesis (Kajiya et al. 2009). In other experimental models of acute cutaneous inflammation, transgenic delivery of VEGF-C and VEGF-D significantly reduced skin inflammation and dermal edema (Huggenberger et al. 2011). Therefore, it appears that VEGF-A and VEGF-C may play opposing roles, the former promoting hyperplastic leaky lymphatic vessels, and the latter improving lymph flow and the resolution of inflammation (Jurisic and Detmar 2009). In peritoneal infection models, experiments with VEGF-C/D blockade and macrophage depletion indicated that the $CD11b^{+}$ macrophage-derived lymphangiogenic factors VEGF-C/D could be major mediators of lipopolysaccharide-induced lymphangiogenesis and lymphatic remodeling (Kim et al. 2009). Under these circumstances, lymphangiogenesis may provide effective conduits for removal of excess interstitial fluid and leaked proteins derived from blood vessels.

Lymphatic obstruction triggering aberrant lymphangiogenesis and lymphatic contractile dysfunction are characteristic features underlying the pathogenesis of chronic inflammatory bowel diseases (IBD) such as Crohn's disease (CD) and ulcerative colitis (UC) (Rahier et al. 2011; von der Weid et al. 2011). Members of the VEGF, bFGF, PDGF-BB, HGF, and angiopoietin families of growth factors have all been implicated in the pathology of IBD (Linares and Gisbert 2011). However, there is still continuing disagreement as to whether lymphatic expansion is protective or detrimental to the progression of IBD. Pedica et al. (2008) suggested that lymphangiogenesis probably contributes to the pathogenesis of CD (Pedica et al. 2008). In contrast, there are indications that the expansion of lymphatic vessel networks in relation to Ang-2 might protect the gut against tissue injury in a dextran sodium sulfate (DSS) model of UC (Ganta et al. 2010).

Lymphangiogenesis is also implicated in several transplant models where blocking lymphangiogenic responses might be beneficial to minimizing transplant rejection and improving allograft survival. It has been shown that VEGFR-3 and VEGF-C-mediated lymphangiogenesis contributes to renal transplant rejection by transporting CCR7-expressing dendritic cells to draining lymph nodes thus eliciting alloantigenic responses (Kerjaschki et al. 2006). Consequently, blocking VEGFR-3-mediated lymphangiogenic events impairs dendritic cell trafficking to draining lymph nodes thus suppressing the adaptive immune response and subsequent rejection of heart, corneal, and pancreatic islet transplants (Chen et al. 2007; Nykanen et al. 2010; Yin et al. 2011). Lymphangiogenesis is also observed in lymph nodes that drain inflamed tissues where it is stimulated by the production of VEGF by the follicular B cells (Angeli and Randolph 2006).

2.3.3 Lymphangiogenesis in Cancer Progression and Metastasis

In many types of cancer such as breast, prostrate, and colon cancer amongst many others, the occurrence of metastasis in the tumor draining lymph node also referred to as the sentinel node is regarded as one of the first clinical signs of malignant tumor cell dissemination. Lymphatic vessels along with blood vessels play a pivotal yet detrimental role in the trafficking of tumor cells, allowing the progression of cancer by providing a route for the metastasis of these tumor cells to distant sites. This is thought to occur in one of two ways, either by invading pre-existing lymphatic vessels or by promoting lymphangiogenesis in response to growth factors (Achen et al. 2005; Tammela and Alitalo 2010). Parameters associated with lymphatic vessels including lymphatic vessel density, lymphovascular invasion, and the expression levels of lymphangiogenic factors represent clinical correlates of tumor staging and the extent of metastasis.

Numerous investigators have reported on active lymphangiogenic mechanisms at the tumor site that employ a plethora of growth factors, cytokines, chemokines, and their receptors and inflammatory agents to promote tumor metastasis and lymphangiogenesis. These include but are not limited to VEGF-C, VEGF-D, VEGFR-3 PDGF-BB, NK-κB, and semaphorins (Albrecht and Christofori 2011; Alitalo 2011; Duong et al. 2012; Tammela and Alitalo 2010). Growth factors, particularly

VEGF-C and VEGF-D, produced by the tumor cells and tumor-associated macrophages facilitate growth of new vessels as well as dilation of existing vessels thus allowing the entry of tumor cells into the vessels (Albrecht and Christofori 2011; Alitalo 2011; Joyce and Pollard 2009). In fact, from a clinical perspective, studies of human cancers have indicated a direct correlation between VEGF-C and VEGF-D expression and lymphatic invasion, metastasis, and poor prognosis of survival (Achen et al. 2005; Alitalo et al. 2005; Tobler and Detmar 2006). In addition, transgenic expression of VEGF-A and the forced expression of PDGF-BB have also been shown to mediate tumor-associated lymphangiogenesis (Bjorndahl et al. 2005b; Cao et al. 2004). While PDGF-BB exerts its effects independent of VEGFR-3, VEGF-A attracts macrophages to the tumor site, which in turn express VEGF-C and VEGF-D thus augmenting tumor-associated lymphangiogenesis. Given the involvement of these growth factors, anti-lymphangiogenic strategies using neutralizing antibodies have been employed to block lymphangiogenesis and tumor metastasis (Chen et al. 2005; He et al. 2002; Lin et al. 2005; Rinderknecht et al. 2010; Roberts et al. 2006). Blocking antibodies against VEGF-C co-receptor, Nrp2 was also successful in inhibiting lymphangiogenesis and tumor metastasis, in part by suppressing LEC migration (Caunt et al. 2008). Tissue remodeling events moderated by the loss of neural cell adhesion molecule resulted in further loss of β1 integrin-mediated tumor cell adhesion, which then upregulated VEGF-C and VEGF-D thereby promoting lymphangiogenesis (Crnic et al. 2004).

There is evidence to indicate that lymphatic vessels within the tumors are structurally disorganized (Achen et al. 2005; Hirakawa et al. 2007; Mandriota et al. 2001; Padera et al. 2002; Skobe et al. 2001; Stacker et al. 2001). In most cases, tumor lymphatics are nonfunctional as they are collapsed due to the high intratumoral pressure to which they are exposed (Padera et al. 2002). Functional lymphatics in the tumor margin are sufficient for lymphatic metastasis, as increases in lymphatic surface area (and thus more opportunity for cancer cell intravasation) have been shown to be accompanied by increases in lymphatic metastasis in VEGF-C-overexpressing tumors, making this an attractive target for controlling tumor growth and metastasis (Padera et al. 2002). From a morphological standpoint, transcription-profiling analyses have determined that tumor-derived LECs are quite different from LECs derived from normal tissue (Clasper et al. 2008; Wu et al. 2010). In essence, lymphatics at the tumor site acquire characteristics prompted by its surrounding environment. They appear to revert to an active lymphangiogenic state similar to developmental lymphangiogenesis. They also derive their cues from the immediate microenvironment and mediate lymphangiogenesis in response to inflammatory signals (Clasper et al. 2008). For example, Cox-2 increases VEGF-C expression in tumor-associated macrophages, thereby augmenting tumor lymphangiogenesis and contributing to metastasis. Therefore, in a mouse model of gastric cancer, Cox-2 inhibition reduced these effects (Iwata et al. 2007).

Tumor lymphangiogenesis is primarily a consequence of new lymphatic sprouts that arise in response to growth factors. To this end, it was shown that VEGF-C stimulated lymphangiogenic sprouting creating intercellular gaps which could enable the entry of tumor cells into the lymphatic vessels (Tammela et al. 2007). In addition to growth factors, hypoxic conditions present at the very core of the tumor further feeds into this aggressive growth of new vessels. Compounding this effect, tumor-infiltrating inflammatory cells such as macrophages express a host of growth factors. Findings from Fischer et al. (2007) demonstrated that anti-PIGF treatment diminished the recruitment of infiltrating macrophages, thereby reducing VEGF-C expression levels and repressing tumor lymphangiogenesis. In addition to sprouting lymphangiogenesis, other cell types such as precursor LECs, hematopoietic stem cells (HSCs), and cells of myeloid–monocyte lineages have been implicated in the formation of new lymphatic vessels (Jiang et al. 2008; Kerjaschki et al. 2006; Maruyama et al. 2005; Religa et al. 2005; Zumsteg et al. 2009). In response to lymphangiogenic factors, lymphatic vessels begin to form sprouts and, driven by LEC proliferation, new lymphatic vessels are formed in the periphery or within the tumor (Albrecht and Christofori 2011). Interestingly, growth factors produced at the tumor site are even capable of initiating lymphangiogenesis in the sentinel lymph node presumably priming the lymph node to create the perfect home for the incoming tumor cells (Harrell et al. 2007; Hirakawa et al. 2007). Metastasis of tumors to distant sites is enabled by tumor-associated lymphangiogenesis, where the lymphatic vessels serve as conduits in the dissemination of the tumor cells. However, several molecules involved in attraction, adhesion, and homing also contribute to the process of tumor metastasis. As an example, secondary lymphoid chemokine also known as CCL21 that is secreted by LECs serves as a chemoattractant for CCR7 expressing melanoma cells, thus allowing higher metastasis to the draining lymph nodes (Shields et al. 2007a). It is also believed that the CCL21 gradient is created by interstitial flow at the tumor site (Shields et al. 2007b). Furthermore, it was shown that CCL21 expression in LECs is upregulated in response to both VEGF-C and interstitial flow (Issa et al. 2009; Miteva et al. 2010). In response to hypoxia, another chemokine implicated in its contribution to tumor metastasis is stromal-cell-derived factor 1 or CXCL12 via its receptors CXCR4 and CXCR7 (Irigoyen et al. 2007). Moreover, enhanced adhesion of the tumor cells to the lymphatic endothelium of lymphatic vessels promotes tumor cell entry. A role for macrophage mannose receptor 1 and CLEVER-1 (stabilin) has been demonstrated in the process of trafficking of tumor cells via tumor-associated lymphatic vessels (Irjala et al. 2003). Importantly, it has been shown that B16 malignant melanoma cells overexpressing CCR7 caused a greater than tenfold increased incidence of regional lymph node metastases after injection into the footpad of mice. Treatment with CCL21-blocking antibodies completely prevented metastasis to the lymph node (Wiley et al. 2001). These findings clearly indicate that the lymphatics may provide the necessary cues whereby cancer cells can harness preexisting molecular mechanisms designed for the physiological immune response to further their progression and metastasis.

2.3.4 Lymphangiogenesis in Lymphovascular Tumors

Lymphangiomas are in essence malformations of the lymphatic vasculature. Most lymphangiomas are benign lesions comprised of thin-walled, cystically dilated vascular channels that are lined by endothelial cells and are filled with proteinaceous lymph. Congenital lymphangiomas are associated with chromosomal abnormalities such as Turner syndrome, whereas acquired lymphangiomas arise due to inflammation, trauma, or lymphatic obstruction. Several hypotheses have been proposed to explain the pathogenesis of lymphangiomas including abnormal budding of the LECs from the cardinal vein, sequestration of primitive lymph tissue during embryologic development, and a failure to separate from the venous vasculature (Wiegand et al. 2008). Other findings indicate that lymphangiomas originate from transformed LECs directed by aberrant regulation of growth factors. A host of lymphangiogenic factors such as Prox1, VEGF-C, and VEGFR-3 results in increased yet impaired lymphangiogenesis culminating in a tumorous proliferative profile.

Lymphangiomas have to be distinguished from other lymphatic tumors like acquired progressive lymphangioma, lymphangiosarcoma, and lymphangiomatosis. Acquired progressive lymphangioma or lymphangioendothelioma is a cutaneous vascular neoplasm comprising of vascular channels that infiltrate the dermis. Lymphangiosarcomas associated with chronic lymphedema post-mastectomy were first described by Stewart and Treves in 1948. It is a rare, malignant, vascular tumor that results as a consequence of postsurgical, filarial, congenital, idiopathic, and traumatic lymphedema (McHaffie et al. 2010). It has been suggested that failed or stagnant lymph transport leads to impaired immune surveillance, making the affected region more prone to the development of vascular neoplasms (Ruocco et al. 2002). Disseminated lymphangiomatosis is a rare disease that is characterized by a proliferation of lymph vessels involving soft tissue, parenchymal organs, dermis, and the skeletal system. Although the molecular mechanisms involved in the pathology of lymphovascular tumors remain vague, there are some indications that modulation of lymphangiogenesis may be of some therapeutic benefit.

Lymphangioleiomyomatosis (LAM) is a benign neoplasm primarily affecting women of child-bearing age. LAM is characterized by the abnormal proliferation and infiltration of smooth muscle-like cells into perivascular spaces and lymphatic vessels. Consequently the pathology results in dysfunctional lymphatics and airway obstruction finally leading to respiratory failure. The muscle-like LAM cells are known to be stimulated by lymphangiogenic growth factors VEGF-C and VEGF-D and several other chemokine receptors. Serum levels of VEGF-D are particularly high in LAM patients and the VEGF-D expression levels correlates with the severity of LAM pathogenesis. Under the influence of these growth factors, LAM cells are able to metastasize via the lymphatic vessels to distant sites, causing further lymphatic obstruction and the progression of fluid accumulation associated with lymphedema. Matrix metalloproteinases (MMPs) are well known for their effects in degrading ECM substrates thereby enabling tumor cell metastasis. Consequently MMPs are also being investigated in relation to LAM pathogenesis. The kinase mammalian target of rapamycin (mTOR) has been implicated in the pathology of LAM via germline mutations in the mTOR repressors, tuberous sclerosis tumor suppressors TSC1 or TSC2. Ongoing clinical trials are evaluating the beneficial effects of rapamycin and mTOR inhibitors such as sirolimus in containing aberrant proliferation of LAM cells (Glasgow et al. 2010).

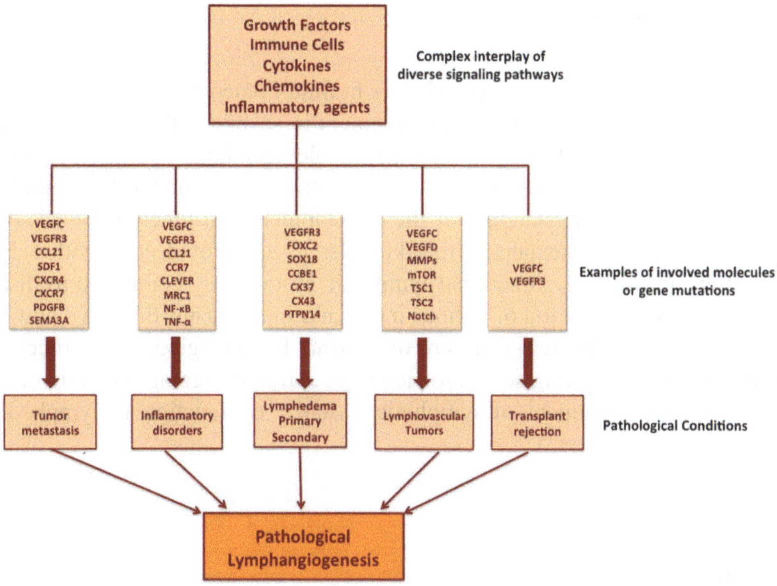

Fig. 2.2 Schematic representation of the involvement of lymphangiogenesis in different pathological conditions. A complex and unregulated interplay of growth factors, immune cells, cytokines, chemokines, and inflammatory agents activate diverse downstream signaling cascades. Some examples are provided of specific molecules that have been reported in literature to be associated with the lymphangiogenic processes underlying various pathological conditions such as tumor metastasis, inflammatory disorders, lymphedema, lymphovascular tumors, and transplant rejection

Kaposi's sarcoma (KS) is a slow progressing tumor caused by human herpes virus 8 (HHV8) or KS-associated herpes virus (KSHV) that manifests as skin lesions. The lesions consist of spindle-like tumor cells, proliferation of vessels rendering them leaky, and extravasated red blood cells (Uldrick and Whitby 2011). KS cells express both BEC and LEC markers (Mesri et al. 2010). Interestingly, the KS virus is capable of reprogramming BECs into a LEC differentiated phenotype and vice versa (Wang et al. 2004). Since KS virus is capable of transforming cells, the involvement of notch signaling pathway, which is also invested with a similar potential, was investigated in relation to KS. It was determined that KSHV could regulate various components of the Notch pathway including Notch receptors (Notch2, Notch3), ligands (Dll1, Dll4, Jagged1), and downstream targets (Hey, Hes) (Liu et al. 2010). Interrupting the notch pathway with a decoy protein in the form of soluble Dll4 (sDll4) inhibited the reprogramming of primary LECs into an invasive mesenchymal phenotype *in vitro*, while reducing tumor growth in vivo (Liu et al. 2010). Since KSHV-infected cells also express VEGFR-3 and VEGF-C, blocking antibodies to VEGFR-3 were employed to demonstrate reduced capillary outgrowth of KSHV-infected LECs (Tvorogov et al. 2010).

An overview of the different pathological conditions involving lymphangiogenesis and some examples of the key regulators promoting this disease state is schematically represented in Fig. 2.2.

2.4 Conclusions

The critical roles played by the lymphatics in fluid homeostasis, lipid metabolism, and immune surveillance and their involvement in several pathological conditions makes it a very valuable target for pharmacological interventions. Progress in advanced imaging techniques and the use of several genetic models has contributed significantly to the understanding of the development and functional basis of the lymphatic vessels. Consequently, this knowledge has provided us with a better appreciation of how a dysfunctional lymphatic system may contribute to disease. Furthermore, the elucidation of various molecular mechanisms that have now been identified to control different stages of the normal lymphangiogenesis process will be instrumental in targeting selective pathways and in designing novel therapeutic strategies for the treatment of lymphatic vascular dysfunction in lymphedema, inflammation, and cancer.

References

Abtahian F, Guerriero A, Sebzda E, Lu MM, Zhou R, Mocsai A, Myers EE, Huang B, Jackson DG, Ferrari VA, Tybulewicz V, Lowell CA, Lepore JJ, Koretzky GA, Kahn ML (2003) Regulation of blood and lymphatic vascular separation by signaling proteins SLP-76 and Syk. Science 299:247–251

Achen MG, Jeltsch M, Kukk E, Makinen T, Vitali A, Wilks AF, Alitalo K, Stacker SA (1998) Vascular endothelial growth factor D (VEGF-D) is a ligand for the tyrosine kinases VEGF receptor 2 (Flk1) and VEGF receptor 3 (Flt4). Proc Natl Acad Sci USA 95:548–553

Achen MG, McColl BK, Stacker SA (2005) Focus on lymphangiogenesis in tumor metastasis. Cancer Cell 7:121–127

Adams RH, Wilkinson GA, Weiss C, Diella F, Gale NW, Deutsch U, Risau W, Klein R (1999) Roles of ephrinB ligands and EphB receptors in cardiovascular development: demarcation of arterial/venous domains, vascular morphogenesis, and sprouting angiogenesis. Genes Dev 13:295–306

Alajati A, Laib AM, Weber H, Boos AM, Bartol A, Ikenberg K, Korff T, Zentgraf H, Obodozie C, Graeser R, Christian S, Finkenzeller G, Stark GB, Heroult M, Augustin HG (2008) Spheroid-based engineering of a human vasculature in mice. Nat Methods 5:439–445

Albrecht I, Christofori G (2011) Molecular mechanisms of lymphangiogenesis in development and cancer. Int J Dev Biol 55:483–494

Alders M, Hogan BM, Gjini E, Salehi F, Al-Gazali L, Hennekam EA, Holmberg EE, Mannens MM, Mulder MF, Offerhaus GJ, Prescott TE, Schroor EJ, Verheij JB, Witte M, Zwijnenburg PJ, Vikkula M, Schulte-Merker S, Hennekam RC (2009) Mutations in CCBE1 cause generalized lymph vessel dysplasia in humans. Nat Genet 41:1272–1274

Alitalo K (2011) The lymphatic vasculature in disease. Nat Med 17:1371–1380

Alitalo K, Tammela T, Petrova TV (2005) Lymphangiogenesis in development and human disease. Nature 438:946–953

Al-Rawi MA, Watkins G, Mansel RE, Jiang WG (2005) The effects of interleukin-7 on the lymphangiogenic properties of human endothelial cells. Int J Oncol 27:721–730

Angeli V, Randolph GJ (2006) Inflammation, lymphatic function, and dendritic cell migration. Lymphat Res Biol 4:217–228

Au AC, Hernandez PA, Lieber E, Nadroo AM, Shen YM, Kelley KA, Gelb BD, Diaz GA (2010) Protein tyrosine phosphatase PTPN14 is a regulator of lymphatic function and choanal development in humans. Am J Hum Genet 87:436–444

Auguste P, Javerzat S, Bikfalvi A (2003) Regulation of vascular development by fibroblast growth factors. Cell Tissue Res 314:157–166

Backhed F, Crawford PA, O'Donnell D, Gordon JI (2007) Postnatal lymphatic partitioning from the blood vasculature in the small intestine requires fasting-induced adipose factor. Proc Natl Acad Sci USA 104:606–611

Baker A, Kim H, Semple JL, Dumont D, Shoichet M, Tobbia D, Johnston M (2010) Experimental assessment of pro-lymphangiogenic growth factors in the treatment of post-surgical lymphedema following lymphadenectomy. Breast Cancer Res 12:R70

Baldwin ME, Halford MM, Roufail S, Williams RA, Hibbs ML, Grail D, Kubo H, Stacker SA, Achen MG (2005) Vascular endothelial growth factor D is dispensable for development of the lymphatic system. Mol Cell Biol 25:2441–2449

Baluk P, Tammela T, Ator E, Lyubynska N, Achen MG, Hicklin DJ, Jeltsch M, Petrova TV, Pytowski B, Stacker SA, Yla-Herttuala S, Jackson DG, Alitalo K, McDonald DM (2005) Pathogenesis of persistent lymphatic vessel hyperplasia in chronic airway inflammation. J Clin Invest 115:247–257

Baluk P, Fuxe J, Hashizume H, Romano T, Lashnits E, Butz S, Vestweber D, Corada M, Molendini C, Dejana E, McDonald DM (2007) Functionally specialized junctions between endothelial cells of lymphatic vessels. J Exp Med 204:2349–2362

Bazigou E, Xie S, Chen C, Weston A, Miura N, Sorokin L, Adams R, Muro AF, Sheppard D, Makinen T (2009) Integrin-alpha9 is required for fibronectin matrix assembly during lymphatic valve morphogenesis. Dev Cell 17:175–186

Bennuru S, Nutman TB (2009) Lymphangiogenesis and lymphatic remodeling induced by filarial parasites: implications for pathogenesis. PLoS Pathog 5:e1000688

Bjorndahl M, Cao R, Nissen LJ, Clasper S, Johnson LA, Xue Y, Zhou Z, Jackson D, Hansen AJ, Cao Y (2005a) Insulin-like growth factors 1 and 2 induce lymphangiogenesis in vivo. Proc Natl Acad Sci USA 102:15593–15598

Bjorndahl MA, Cao R, Burton JB, Brakenhielm E, Religa P, Galter D, Wu L, Cao Y (2005b) Vascular endothelial growth factor-a promotes peritumoral lymphangiogenesis and lymphatic metastasis. Cancer Res 65:9261–9268

Bollinger A, Isenring G, Franzeck UK, Brunner U (1983) Aplasia of superficial lymphatic capillaries in hereditary and connatal lymphedema (Milroy's disease). Lymphology 16:27–30

Breiteneder-Geleff S, Soleiman A, Kowalski H, Horvat R, Amann G, Kriehuber E, Diem K, Weninger W, Tschachler E, Alitalo K, Kerjaschki D (1999) Angiosarcomas express mixed endothelial phenotypes of blood and lymphatic capillaries: podoplanin as a specific marker for lymphatic endothelium. Am J Pathol 154:385–394

Bruyere F, Noel A (2010) Lymphangiogenesis: in vitro and in vivo models. FASEB J 24:8–21

Bruyere F, Melen-Lamalle L, Blacher S, Roland G, Thiry M, Moons L, Frankenne F, Carmeliet P, Alitalo K, Libert C, Sleeman JP, Foidart JM, Noel A (2008) Modeling lymphangiogenesis in a three-dimensional culture system. Nat Methods 5:431–437

Byzova TV, Goldman CK, Jankau J, Chen J, Cabrera G, Achen MG, Stacker SA, Carnevale KA, Siemionow M, Deitcher SR, DiCorleto PE (2002) Adenovirus encoding vascular endothelial growth factor-D induces tissue-specific vascular patterns in vivo. Blood 99:4434–4442

Cao R, Bjorndahl MA, Religa P, Clasper S, Garvin S, Galter D, Meister B, Ikomi F, Tritsaris K, Dissing S, Ohhashi T, Jackson DG, Cao Y (2004) PDGF-BB induces intratumoral lymphangiogenesis and promotes lymphatic metastasis. Cancer Cell 6:333–345

Carmeliet P, Ferreira V, Breier G, Pollefeyt S, Kieckens L, Gertsenstein M, Fahrig M, Vandenhoeck A, Harpal K, Eberhardt C, Declercq C, Pawling J, Moons L, Collen D, Risau W, Nagy A (1996) Abnormal blood vessel development and lethality in embryos lacking a single VEGF allele. Nature 380:435–439

Casley-Smith JR (1974) The lymphatic system in inflammation. In: Zweifach BW, Grant L, Mcclusky RT (eds) The inflammatory process. Academic, New York, pp 161–204

Caunt M, Mak J, Liang WC, Stawicki S, Pan Q, Tong RK, Kowalski J, Ho C, Reslan HB, Ross J, Berry L, Kasman I, Zlot C, Cheng Z, Le Couter J, Filvaroff EH, Plowman G, Peale F, French D, Carano R, Koch AW, Wu Y, Watts RJ, Tessier-Lavigne M, Bagri A (2008) Blocking neuropilin-2 function inhibits tumor cell metastasis. Cancer Cell 13:331–342

Chen Z, Varney ML, Backora MW, Cowan K, Solheim JC, Talmadge JE, Singh RK (2005) Down-regulation of vascular endothelial cell growth factor-C expression using small interfering RNA vectors in mammary tumors inhibits tumor lymphangiogenesis and spontaneous metastasis and enhances survival. Cancer Res 65:9004–9011

Chen L, Hamrah P, Cursiefen C, Zhang Q, Pytowski B, Streilein JW, Dana MR (2007) Vascular endothelial growth factor receptor-3 mediates induction of corneal alloimmunity. 2004. Ocul Immunol Inflamm 15:275–278

Clasper S, Royston D, Baban D, Cao Y, Ewers S, Butz S, Vestweber D, Jackson DG (2008) A novel gene expression profile in lymphatics associated with tumor growth and nodal metastasis. Cancer Res 68:7293–7303

Connell FC, Ostergaard P, Carver C, Brice G, Williams N, Mansour S, Mortimer PS, Jeffery S (2009) Analysis of the coding regions of VEGFR3 and VEGFC in Milroy disease and other primary lymphoedemas. Hum Genet 124:625–631

Crnic I, Strittmatter K, Cavallaro U, Kopfstein L, Jussila L, Alitalo K, Christofori G (2004) Loss of neural cell adhesion molecule induces tumor metastasis by up-regulating lymphangiogenesis. Cancer Res 64:8630–8638

Cueni LN, Detmar M (2008) The lymphatic system in health and disease. Lymphat Res Biol 6:109–122

Dagenais SL, Hartsough RL, Erickson RP, Witte MH, Butler MG, Glover TW (2004) Foxc2 is expressed in developing lymphatic vessels and other tissues associated with lymphedema-distichiasis syndrome. Gene Expr Patterns 4:611–619

Danussi C, Spessotto P, Petrucco A, Wassermann B, Sabatelli P, Montesi M, Doliana R, Bressan G, Colombatti A (2008) Emilin1 deficiency causes structural and functional defects of lymphatic vasculature. Mol Cell Biol 28:4026–4039

Dejana E, Orsenigo F, Molendini C, Baluk P, McDonald DM (2009) Organization and signaling of endothelial cell-to-cell junctions in various regions of the blood and lymphatic vascular trees. Cell Tissue Res 335:17–25

Dellinger M, Hunter R, Bernas M, Gale N, Yancopoulos G, Erickson R, Witte M (2008) Defective remodeling and maturation of the lymphatic vasculature in angiopoietin-2 deficient mice. Dev Biol 319:309–320

Dreyer G, Noroes J, Figueredo-Silva J, Piessens WF (2000) Pathogenesis of lymphatic disease in bancroftian filariasis: a clinical perspective. Parasitol Today 16:544–548

Dumont DJ, Jussila L, Taipale J, Lymboussaki A, Mustonen T, Pajusola K, Breitman M, Alitalo K (1998) Cardiovascular failure in mouse embryos deficient in VEGF receptor-3. Science 282:946–949

Duong T, Koopman P, Francois M (2012) Tumor lymphangiogenesis as a potential therapeutic target. J Oncol 2012:204946

Enholm B, Karpanen T, Jeltsch M, Kubo H, Stenback F, Prevo R, Jackson DG, Yla-Herttuala S, Alitalo K (2001) Adenoviral expression of vascular endothelial growth factor-C induces lymphangiogenesis in the skin. Circ Res 88:623–629

Fang J, Dagenais SL, Erickson RP, Arlt MF, Glynn MW, Gorski JL, Seaver LH, Glover TW (2000) Mutations in FOXC2 (MFH-1), a forkhead family transcription factor, are responsible for the hereditary lymphedema-distichiasis syndrome. Am J Hum Genet 67:1382–1388

Ferrell RE, Baty CJ, Kimak MA, Karlsson JM, Lawrence EC, Franke-Snyder M, Meriney SD, Feingold E, Finegold DN (2010) GJC2 missense mutations cause human lymphedema. Am J Hum Genet 86:943–948

Finegold DN, Kimak MA, Lawrence EC, Levinson KL, Cherniske EM, Pober BR, Dunlap JW, Ferrell RE (2001) Truncating mutations in FOXC2 cause multiple lymphedema syndromes. Hum Mol Genet 10:1185–1189

Fischer C, Jonckx B, Mazzone M, Zacchigna S, Loges S, Pattarini L, Chorianopoulos E, Liesenborghs L, Koch M, De Mol M, Autiero M, Wyns S, Plaisance S, Moons L, van Rooijen N, Giacca M, Stassen JM, Dewerchin M, Collen D, Carmeliet P (2007) Anti-PlGF inhibits growth of VEGF(R)-inhibitor-resistant tumors without affecting healthy vessels. Cell 131:463–475

Flister MJ, Wilber A, Hall KL, Iwata C, Miyazono K, Nisato RE, Pepper MS, Zawieja DC, Ran S (2010) Inflammation induces lymphangiogenesis through up-regulation of VEGFR-3 mediated by NF-kappaB and Prox1. Blood 115:418–429

Foskett AM, Ezekiel UR, Trzeciakowski JP, Zawieja DC, Muthuchamy M (2011) Hypoxia and extracellular matrix proteins influence angiogenesis and lymphangiogenesis in mouse embryoid bodies. Front Physiol 2:103

Francois M, Caprini A, Hosking B, Orsenigo F, Wilhelm D, Browne C, Paavonen K, Karnezis T, Shayan R, Downes M, Davidson T, Tutt D, Cheah KS, Stacker SA, Muscat GE, Achen MG, Dejana E, Koopman P (2008) Sox18 induces development of the lymphatic vasculature in mice. Nature 456:643–647

Gale NW, Baluk P, Pan L, Kwan M, Holash J, DeChiara TM, McDonald DM, Yancopoulos GD (2001) Ephrin-B2 selectively marks arterial vessels and neovascularization sites in the adult, with expression in both endothelial and smooth-muscle cells. Dev Biol 230:151–160

Gale NW, Thurston G, Hackett SF, Renard R, Wang Q, McClain J, Martin C, Witte C, Witte MH, Jackson D, Suri C, Campochiaro PA, Wiegand SJ, Yancopoulos GD (2002) Angiopoietin-2 is required for postnatal angiogenesis and lymphatic patterning, and only the latter role is rescued by angiopoietin-1. Dev Cell 3:411–423

Gale NW, Prevo R, Espinosa J, Ferguson DJ, Dominguez MG, Yancopoulos GD, Thurston G, Jackson DG (2007) Normal lymphatic development and function in mice deficient for the lymphatic hyaluronan receptor LYVE-1. Mol Cell Biol 27:595–604

Ganta VC, Cromer W, Mills GL, Traylor J, Jennings M, Daley S, Clark B, Mathis JM, Bernas M, Boktor M, Jordan P, Witte M, Alexander JS (2010) Angiopoietin-2 in experimental colitis. Inflamm Bowel Dis 16:1029–1039

Gashev AA (2002) Physiologic aspects of lymphatic contractile function: current perspectives. Ann N Y Acad Sci 979:178–187; discussion 188–196

Gashev AA, Li J, Muthuchamy M, Zawieja DC (2012) Adenovirus-mediated gene transfection in the isolated lymphatic vessels. Methods Mol Biol 843:199–204

Glasgow CG, Steagall WK, Taveira-Dasilva A, Pacheco-Rodriguez G, Cai X, El-Chemaly S, Moses M, Darling T, Moss J (2010) Lymphangioleiomyomatosis (LAM): molecular insights lead to targeted therapies. Respir Med 104(suppl 1):S45–S58

Gnepp DR (1976) The bicuspid nature of the valves of the peripheral collecting lymphatic vessels of the dog. Lymphology 9:75–77

Gnepp DR (1984) Lymphatics. In: Staub NC, Taylor AE (eds) Edema. Raven Press, New York, NY, pp 263–298

Gnepp DR, Chandler W (1985) Tissue culture of human and canine thoracic duct endothelium. In Vitro Cell Dev Biol 21:200–206

Gnepp DR, Green FH (1980) Scanning electron microscopic study of canine lymphatic vessels and their valves. Lymphology 13:91–99

Groger M, Loewe R, Holnthoner W, Embacher R, Pillinger M, Herron GS, Wolff K, Petzelbauer P (2004) IL-3 induces expression of lymphatic markers Prox-1 and podoplanin in human endothelial cells. J Immunol 173:7161–7169

Hamada K, Oike Y, Takakura N, Ito Y, Jussila L, Dumont DJ, Alitalo K, Suda T (2000) VEGF-C signaling pathways through VEGFR-2 and VEGFR-3 in vasculoangiogenesis and hematopoiesis. Blood 96:3793–3800

Harrell MI, Iritani BM, Ruddell A (2007) Tumor-induced sentinel lymph node lymphangiogenesis and increased lymph flow precede melanoma metastasis. Am J Pathol 170:774–786

Harvey NL, Srinivasan RS, Dillard ME, Johnson NC, Witte MH, Boyd K, Sleeman MW, Oliver G (2005) Lymphatic vascular defects promoted by Prox1 haploinsufficiency cause adult-onset obesity. Nat Genet 37:1072–1081

Hayes H, Kossmann E, Wilson E, Meininger C, Zawieja D (2003) Development and characterization of endothelial cells from rat microlymphatics. Lymphat Res Biol 1:101–119

He Y, Kozaki K, Karpanen T, Koshikawa K, Yla-Herttuala S, Takahashi T, Alitalo K (2002) Suppression of tumor lymphangiogenesis and lymph node metastasis by blocking vascular endothelial growth factor receptor 3 signaling. J Natl Cancer Inst 94:819–825

Helm CL, Fleury ME, Zisch AH, Boschetti F, Swartz MA (2005) Synergy between interstitial flow and VEGF directs capillary morphogenesis in vitro through a gradient amplification mechanism. Proc Natl Acad Sci USA 102:15779–15784

Helm CL, Zisch A, Swartz MA (2007) Engineered blood and lymphatic capillaries in 3-D VEGF-fibrin-collagen matrices with interstitial flow. Biotechnol Bioeng 96:167–176

Hirakawa S, Hong YK, Harvey N, Schacht V, Matsuda K, Libermann T, Detmar M (2003) Identification of vascular lineage-specific genes by transcriptional profiling of isolated blood vascular and lymphatic endothelial cells. Am J Pathol 162:575–586

Hirakawa S, Brown LF, Kodama S, Paavonen K, Alitalo K, Detmar M (2007) VEGF-C-induced lymphangiogenesis in sentinel lymph nodes promotes tumor metastasis to distant sites. Blood 109:1010–1017

Hirashima M, Sano K, Morisada T, Murakami K, Rossant J, Suda T (2008) Lymphatic vessel assembly is impaired in Aspp1-deficient mouse embryos. Dev Biol 316:149–159

Hong YK, Harvey N, Noh YH, Schacht V, Hirakawa S, Detmar M, Oliver G (2002) Prox1 is a master control gene in the program specifying lymphatic endothelial cell fate. Dev Dyn 225:351–357

Hong YK, Foreman K, Shin JW, Hirakawa S, Curry CL, Sage DR, Libermann T, Dezube BJ, Fingeroth JD, Detmar M (2004) Lymphatic reprogramming of blood vascular endothelium by Kaposi sarcoma-associated herpesvirus. Nat Genet 36:683–685

Huggenberger R, Siddiqui SS, Brander D, Ullmann S, Zimmermann K, Antsiferova M, Werner S, Alitalo K, Detmar M (2011) An important role of lymphatic vessel activation in limiting acute inflammation. Blood 117:4667–4678

Huntington GS, McClure CFW (1910) The anatomy and development of the jugular lymph sac in the domestic cat (Felis domestica). Am J Anat 10:177–311

Ichise H, Ichise T, Ohtani O, Yoshida N (2009) Phospholipase Cgamma2 is necessary for separation of blood and lymphatic vasculature in mice. Development 136:191–195

Irigoyen M, Anso E, Martinez E, Garayoa M, Martinez-Irujo JJ, Rouzaut A (2007) Hypoxia alters the adhesive properties of lymphatic endothelial cells. A transcriptional and functional study. Biochim Biophys Acta 1773:880–890

Irjala H, Alanen K, Grenman R, Heikkila P, Joensuu H, Jalkanen S (2003) Mannose receptor (MR) and common lymphatic endothelial and vascular endothelial receptor (CLEVER)-1 direct the binding of cancer cells to the lymph vessel endothelium. Cancer Res 63:4671–4676

Irrthum A, Devriendt K, Chitayat D, Matthijs G, Glade C, Steijlen PM, Fryns JP, Van Steensel MA, Vikkula M (2003) Mutations in the transcription factor gene SOX18 underlie recessive and dominant forms of hypotrichosis-lymphedema-telangiectasia. Am J Hum Genet 72:1470–1478

Isogai S, Hitomi J, Yaniv K, Weinstein BM (2009) Zebrafish as a new animal model to study lymphangiogenesis. Anat Sci Int 84:102–111

Issa A, Le TX, Shoushtari AN, Shields JD, Swartz MA (2009) Vascular endothelial growth factor-C and C-C chemokine receptor 7 in tumor cell-lymphatic cross-talk promote invasive phenotype. Cancer Res 69:349–357

Iwata C, Kano MR, Komuro A, Oka M, Kiyono K, Johansson E, Morishita Y, Yashiro M, Hirakawa K, Kaminishi M, Miyazono K (2007) Inhibition of cyclooxygenase-2 suppresses lymph node metastasis via reduction of lymphangiogenesis. Cancer Res 67:10181–10189

Jeltsch M, Kaipainen A, Joukov V, Meng X, Lakso M, Rauvala H, Swartz M, Fukumura D, Jain RK, Alitalo K (1997) Hyperplasia of lymphatic vessels in VEGF-C transgenic mice. Science 276:1423–1425

Jiang S, Bailey AS, Goldman DC, Swain JR, Wong MH, Streeter PR, Fleming WH (2008) Hematopoietic stem cells contribute to lymphatic endothelium. PLoS One 3:e3812

Johnson LA, Clasper S, Holt AP, Lalor PF, Baban D, Jackson DG (2006) An inflammation-induced mechanism for leukocyte transmigration across lymphatic vessel endothelium. J Exp Med 203:2763–2777

Joukov V, Pajusola K, Kaipainen A, Chilov D, Lahtinen I, Kukk E, Saksela O, Kalkkinen N, Alitalo K (1996) A novel vascular endothelial growth factor, VEGF-C, is a ligand for the Flt4 (VEGFR-3) and KDR (VEGFR-2) receptor tyrosine kinases. EMBO J 15:1751

Joyce JA, Pollard JW (2009) Microenvironmental regulation of metastasis. Nat Rev Cancer 9:239–252

Jurisic G, Detmar M (2009) Lymphatic endothelium in health and disease. Cell Tissue Res 335:97–108

Kaipainen A, Korhonen J, Mustonen T, van Hinsbergh VW, Fang GH, Dumont D, Breitman M, Alitalo K (1995) Expression of the fms-like tyrosine kinase 4 gene becomes restricted to lymphatic endothelium during development. Proc Natl Acad Sci USA 92:3566–3570

Kajiya K, Detmar M (2006) An important role of lymphatic vessels in the control of UVB-induced edema formation and inflammation. J Invest Dermatol 126:919–921

Kajiya K, Hirakawa S, Ma B, Drinnenberg I, Detmar M (2005) Hepatocyte growth factor promotes lymphatic vessel formation and function. EMBO J 24:2885–2895

Kajiya K, Hirakawa S, Detmar M (2006) Vascular endothelial growth factor-A mediates ultraviolet B-induced impairment of lymphatic vessel function. Am J Pathol 169:1496–1503

Kajiya K, Sawane M, Huggenberger R, Detmar M (2009) Activation of the VEGFR-3 pathway by VEGF-C attenuates UVB-induced edema formation and skin inflammation by promoting lymphangiogenesis. J Invest Dermatol 129:1292–1298

Kanady JD, Dellinger MT, Munger SJ, Witte MH, Simon AM (2011) Connexin37 and Connexin43 deficiencies in mice disrupt lymphatic valve development and result in lymphatic disorders including lymphedema and chylothorax. Dev Biol 354:253–266

Kang S, Lee SP, Kim KE, Kim HZ, Memet S, Koh GY (2009) Toll-like receptor 4 in lymphatic endothelial cells contributes to LPS-induced lymphangiogenesis by chemotactic recruitment of macrophages. Blood 113:2605–2613

Karkkainen MJ, Ferrell RE, Lawrence EC, Kimak MA, Levinson KL, McTigue MA, Alitalo K, Finegold DN (2000) Missense mutations interfere with VEGFR-3 signalling in primary lymphoedema. Nat Genet 25:153–159

Karkkainen MJ, Saaristo A, Jussila L, Karila KA, Lawrence EC, Pajusola K, Bueler H, Eichmann A, Kauppinen R, Kettunen MI, Yla-Herttuala S, Finegold DN, Ferrell RE, Alitalo K (2001) A model for gene therapy of human hereditary lymphedema. Proc Natl Acad Sci USA 98:12677–12682

Karkkainen MJ, Haiko P, Sainio K, Partanen J, Taipale J, Petrova TV, Jeltsch M, Jackson DG, Talikka M, Rauvala H, Betsholtz C, Alitalo K (2004) Vascular endothelial growth factor C is required for sprouting of the first lymphatic vessels from embryonic veins. Nat Immunol 5:74–80

Karpanen T, Alitalo K (2008) Molecular biology and pathology of lymphangiogenesis. Annu Rev Pathol 3:367–397

Kawai Y, Hosaka K, Kaidoh M, Minami T, Kodama T, Ohhashi T (2008) Heterogeneity in immunohistochemical, genomic, and biological properties of human lymphatic endothelial cells between initial and collecting lymph vessels. Lymphat Res Biol 6:15–27

Kerjaschki D (2005) The crucial role of macrophages in lymphangiogenesis. J Clin Invest 115:2316–2319

Kerjaschki D, Huttary N, Raab I, Regele H, Bojarski-Nagy K, Bartel G, Krober SM, Greinix H, Rosenmaier A, Karlhofer F, Wick N, Mazal PR (2006) Lymphatic endothelial progenitor cells contribute to de novo lymphangiogenesis in human renal transplants. Nat Med 12:230–234

Kim KE, Koh YJ, Jeon BH, Jang C, Han J, Kataru RP, Schwendener RA, Kim JM, Koh GY (2009) Role of CD11b+ macrophages in intraperitoneal lipopolysaccharide-induced aberrant lymphangiogenesis and lymphatic function in the diaphragm. Am J Pathol 175:1733–1745

Kreuger J, Nilsson I, Kerjaschki D, Petrova T, Alitalo K, Claesson-Welsh L (2006) Early lymph vessel development from embryonic stem cells. Arterioscler Thromb Vasc Biol 26:1073–1078

Kriehuber E, Breiteneder-Geleff S, Groeger M, Soleiman A, Schoppmann SF, Stingl G, Kerjaschki D, Maurer D (2001) Isolation and characterization of dermal lymphatic and blood endothelial cells reveal stable and functionally specialized cell lineages. J Exp Med 194:797–808

Kubo H, Cao R, Brakenhielm E, Makinen T, Cao Y, Alitalo K (2002) Blockade of vascular endothelial growth factor receptor-3 signaling inhibits fibroblast growth factor-2-induced lymphangiogenesis in mouse cornea. Proc Natl Acad Sci USA 99:8868–8873

Kuchler AM, Gjini E, Peterson-Maduro J, Cancilla B, Wolburg H, Schulte-Merker S (2006) Development of the zebrafish lymphatic system requires VEGFC signaling. Curr Biol 16:1244–1248

Lahteenvuo M, Honkonen K, Tervala T, Tammela T, Suominen E, Lahteenvuo J, Kholova I, Alitalo K, Yla-Herttuala S, Saaristo A (2011) Growth factor therapy and autologous lymph node transfer in lymphedema. Circulation 123:613–620

Leak LV, Burke JF (1966) Fine structure of the lymphatic capillary and the adjoining connective tissue area. Am J Anat 118:785–809

Leak LV, Burke JF (1968) Ultrastructural studies on the lymphatic anchoring filaments. J Cell Biol 36:129–149

Leak LV, Jones M (1994) Lymphangiogenesis in vitro: formation of lymphatic capillary-like channels from confluent monolayers of lymphatic endothelial cells. In Vitro Cell Dev Biol Anim 30A:512–518

Lee S, Kang J, Yoo J, Ganesan SK, Cook SC, Aguilar B, Ramu S, Lee J, Hong YK (2009) Prox1 physically and functionally interacts with COUP-TFII to specify lymphatic endothelial cell fate. Blood 113:1856–1859

Liersch R, Nay F, Lu L, Detmar M (2006) Induction of lymphatic endothelial cell differentiation in embryoid bodies. Blood 107:1214–1216

Lin J, Lalani AS, Harding TC, Gonzalez M, Wu WW, Luan B, Tu GH, Koprivnikar K, VanRoey MJ, He Y, Alitalo K, Jooss K (2005) Inhibition of lymphogenous metastasis using adeno-associated virus-mediated gene transfer of a soluble VEGFR-3 decoy receptor. Cancer Res 65:6901–6909

Lin FJ, Chen X, Qin J, Hong YK, Tsai MJ, Tsai SY (2010) Direct transcriptional regulation of neuropilin-2 by COUP-TFII modulates multiple steps in murine lymphatic vessel development. J Clin Invest 120:1694–1707

Linares PM, Gisbert JP (2011) Role of growth factors in the development of lymphangiogenesis driven by inflammatory bowel disease: a review. Inflamm Bowel Dis 17:1814–1821

Liu R, Li X, Tulpule A, Zhou Y, Scehnet JS, Zhang S, Lee JS, Chaudhary PM, Jung J, Gill PS (2010) KSHV-induced notch components render endothelial and mural cell characteristics and cell survival. Blood 115:887–895

Lohela M, Bry M, Tammela T, Alitalo K (2009) VEGFs and receptors involved in angiogenesis versus lymphangiogenesis. Curr Opin Cell Biol 21:154–165

Luong MX, Tam J, Lin Q, Hagendoorn J, Moore KJ, Padera TP, Seed B, Fukumura D, Kucherlapati R, Jain RK (2009) Lack of lymphatic vessel phenotype in LYVE-1/CD44 double knockout mice. J Cell Physiol 219:430–437

Maisonpierre PC, Suri C, Jones PF, Bartunkova S, Wiegand SJ, Radziejewski C, Compton D, McClain J, Aldrich TH, Papadopoulos N, Daly TJ, Davis S, Sato TN, Yancopoulos GD (1997) Angiopoietin-2, a natural antagonist for Tie2 that disrupts in vivo angiogenesis. Science 277:55–60

Makinen T, Adams RH, Bailey J, Lu Q, Ziemiecki A, Alitalo K, Klein R, Wilkinson GA (2005) PDZ interaction site in ephrinB2 is required for the remodeling of lymphatic vasculature. Genes Dev 19:397–410

Makinen T, Norrmen C, Petrova TV (2007) Molecular mechanisms of lymphatic vascular development. Cell Mol Life Sci 64:1915–1929

Mandriota SJ, Jussila L, Jeltsch M, Compagni A, Baetens D, Prevo R, Banerji S, Huarte J, Montesano R, Jackson DG, Orci L, Alitalo K, Christofori G, Pepper MS (2001) Vascular endothelial growth factor-C-mediated lymphangiogenesis promotes tumour metastasis. EMBO J 20:672–682

Maruyama K, Ii M, Cursiefen C, Jackson DG, Keino H, Tomita M, Van Rooijen N, Takenaka H, D'Amore PA, Stein-Streilein J, Losordo DW, Streilein JW (2005) Inflammation-induced lymphangiogenesis in the cornea arises from CD11b-positive macrophages. J Clin Invest 115:2363–2372

McHaffie DR, Kozak KR, Warner TF, Cho CS, Heiner JP, Attia S (2010) Stewart-Treves syndrome of the lower extremity. J Clin Oncol 28:e351–e352

Mellor RH, Brice G, Stanton AW, French J, Smith A, Jeffery S, Levick JR, Burnand KG, Mortimer PS, Lymphoedema Research Consortium (2007) Mutations in FOXC2 are strongly associated with primary valve failure in veins of the lower limb. Circulation 115:1912–1920

Mellor RH, Hubert CE, Stanton AW, Tate N, Akhras V, Smith A, Burnand KG, Jeffery S, Makinen T, Levick JR, Mortimer PS (2010) Lymphatic dysfunction, not aplasia, underlies Milroy disease. Microcirculation 17:281–296

Mesri EA, Cesarman E, Boshoff C (2010) Kaposi's sarcoma and its associated herpesvirus. Nat Rev Cancer 10:707–719

Miteva DO, Rutkowski JM, Dixon JB, Kilarski W, Shields JD, Swartz MA (2010) Transmural flow modulates cell and fluid transport functions of lymphatic endothelium. Circ Res 106:920–931

Mizuno R, Yokoyama Y, Ono N, Ikomi F, Ohhashi T (2003) Establishment of rat lymphatic endothelial cell line. Microcirculation 10:127–131

Mouta C, Heroult M (2003) Inflammatory triggers of lymphangiogenesis. Lymphat Res Biol 1:201–218

Murdaca G, Cagnati P, Gulli R, Spano F, Puppo F, Campisi C, Boccardo F (2012) Current views on diagnostic approach and treatment of lymphedema. Am J Med 125:134–140

Neufeld G, Cohen T, Shraga N, Lange T, Kessler O, Herzog Y (2002) The neuropilins: multifunctional semaphorin and VEGF receptors that modulate axon guidance and angiogenesis. Trends Cardiovasc Med 12:13–19

Ng CP, Helm CL, Swartz MA (2004) Interstitial flow differentially stimulates blood and lymphatic endothelial cell morphogenesis in vitro. Microvasc Res 68:258–264

Nilsson I, Rolny C, Wu Y, Pytowski B, Hicklin D, Alitalo K, Claesson-Welsh L, Wennstrom S (2004) Vascular endothelial growth factor receptor-3 in hypoxia-induced vascular development. FASEB J 18:1507–1515

Norrmen C, Ivanov KI, Cheng J, Zangger N, Delorenzi M, Jaquet M, Miura N, Puolakkainen P, Horsley V, Hu J, Augustin HG, Yla-Herttuala S, Alitalo K, Petrova TV (2009) FOXC2 controls formation and maturation of lymphatic collecting vessels through cooperation with NFATc1. J Cell Biol 185:439–457

Norrmen C, Tammela T, Petrova TV, Alitalo K (2011) Biological basis of therapeutic lymphangiogenesis. Circulation 123:1335–1351

Ny A, Koch M, Schneider M, Neven E, Tong RT, Maity S, Fischer C, Plaisance S, Lambrechts D, Heligon C, Terclavers S, Ciesiolka M, Kalin R, Man WY, Senn I, Wyns S, Lupu F, Brandli A, Vleminckx K, Collen D, Dewerchin M, Conway EM, Moons L, Jain RK, Carmeliet P (2005) A genetic Xenopus laevis tadpole model to study lymphangiogenesis. Nat Med 11:998–1004

Ny A, Autiero M, Carmeliet P (2006) Zebrafish and Xenopus tadpoles: small animal models to study angiogenesis and lymphangiogenesis. Exp Cell Res 312:684–693

Nykanen AI, Sandelin H, Krebs R, Keranen MA, Tuuminen R, Karpanen T, Wu Y, Pytowski B, Koskinen PK, Yla-Herttuala S, Alitalo K, Lemstrom KB (2010) Targeting lymphatic vessel activation and CCL21 production by vascular endothelial growth factor receptor-3 inhibition has novel immunomodulatory and antiarteriosclerotic effects in cardiac allografts. Circulation 121:1413–1422

Oh SJ, Jeltsch MM, Birkenhager R, McCarthy JE, Weich HA, Christ B, Alitalo K, Wilting J (1997) VEGF and VEGF-C: specific induction of angiogenesis and lymphangiogenesis in the differentiated avian chorioallantoic membrane. Dev Biol 188:96–109

Ohtani Y, Ohtani O (2001) Postnatal development of lymphatic vessels and their smooth muscle cells in the rat diaphragm: a confocal microscopic study. Arch Histol Cytol 64:513–522

Okazaki T, Ni A, Ayeni OA, Baluk P, Yao LC, Vossmeyer D, Zischinsky G, Zahn G, Knolle J, Christner C, McDonald DM (2009) Alpha5beta1 Integrin blockade inhibits lymphangiogenesis in airway inflammation. Am J Pathol 174:2378–2387

Oliver G (2004) Lymphatic vasculature development. Nat Rev Immunol 4:35–45

Oliver G, Alitalo K (2005) The lymphatic vasculature: recent progress and paradigms. Annu Rev Cell Dev Biol 21:457–483

Oliver G, Detmar M (2002) The rediscovery of the lymphatic system: old and new insights into the development and biological function of the lymphatic vasculature. Genes Dev 16:773–783

Oliver G, Srinivasan RS (2008) Lymphatic vasculature development: current concepts. Ann N Y Acad Sci 1131:75–81

Padera TP, Kadambi A, di Tomaso E, Carreira CM, Brown EB, Boucher Y, Choi NC, Mathisen D, Wain J, Mark EJ, Munn LL, Jain RK (2002) Lymphatic metastasis in the absence of functional intratumor lymphatics. Science 296:1883–1886

Pedica F, Ligorio C, Tonelli P, Bartolini S, Baccarini P (2008) Lymphangiogenesis in Crohn's disease: an immunohistochemical study using monoclonal antibody D2-40. Virchows Arch 452:57–63

Petrenko VM, Gashev AA (2008) Observations on the prenatal development of human lymphatic vessels with focus on basic structural elements of lymph flow. Lymphat Res Biol 6:89–95

Petrova TV, Makinen T, Makela TP, Saarela J, Virtanen I, Ferrell RE, Finegold DN, Kerjaschki D, Yla-Herttuala S, Alitalo K (2002) Lymphatic endothelial reprogramming of vascular endothelial cells by the Prox-1 homeobox transcription factor. EMBO J 21:4593–4599

Petrova TV, Karpanen T, Norrmen C, Mellor R, Tamakoshi T, Finegold D, Ferrell R, Kerjaschki D, Mortimer P, Yla-Herttuala S, Miura N, Alitalo K (2004) Defective valves and abnormal mural cell recruitment underlie lymphatic vascular failure in lymphedema distichiasis. Nat Med 10:974–981

Pfarr KM, Debrah AY, Specht S, Hoerauf A (2009) Filariasis and lymphoedema. Parasite Immunol 31:664–672

Podgrabinska S, Braun P, Velasco P, Kloos B, Pepper MS, Skobe M (2002) Molecular characterization of lymphatic endothelial cells. Proc Natl Acad Sci USA 99:16069–16074

Polzer K, Baeten D, Soleiman A, Distler J, Gerlag DM, Tak PP, Schett G, Zwerina J (2008) Tumour necrosis factor blockade increases lymphangiogenesis in murine and human arthritic joints. Ann Rheum Dis 67:1610–1616

Rahier JF, De Beauce S, Dubuquoy L, Erdual E, Colombel JF, Jouret-Mourin A, Geboes K, Desreumaux P (2011) Increased lymphatic vessel density and lymphangiogenesis in inflammatory bowel disease. Aliment Pharmacol Ther 34:533–543

Randolph GJ, Angeli V, Swartz MA (2005) Dendritic-cell trafficking to lymph nodes through lymphatic vessels. Nat Rev Immunol 5:617–628

Ranger AM, Grusby MJ, Hodge MR, Gravallese EM, de la Brousse FC, Hoey T, Mickanin C, Baldwin HS, Glimcher LH (1998) The transcription factor NF-ATc is essential for cardiac valve formation. Nature 392:186–190

Religa P, Cao R, Bjorndahl M, Zhou Z, Zhu Z, Cao Y (2005) Presence of bone marrow-derived circulating progenitor endothelial cells in the newly formed lymphatic vessels. Blood 106:4184–4190

Rinderknecht M, Villa A, Ballmer-Hofer K, Neri D, Detmar M (2010) Phage-derived fully human monoclonal antibody fragments to human vascular endothelial growth factor-C block its interaction with VEGF receptor-2 and 3. PLoS One 5:e11941

Rissanen TT, Markkanen JE, Gruchala M, Heikura T, Puranen A, Kettunen MI, Kholova I, Kauppinen RA, Achen MG, Stacker SA, Alitalo K, Yla-Herttuala S (2003) VEGF-D is the strongest angiogenic and lymphangiogenic effector among VEGFs delivered into skeletal muscle via adenoviruses. Circ Res 92:1098–1106

Ristimaki A, Narko K, Enholm B, Joukov V, Alitalo K (1998) Proinflammatory cytokines regulate expression of the lymphatic endothelial mitogen vascular endothelial growth factor-C. J Biol Chem 273:8413–8418

Roberts N, Kloos B, Cassella M, Podgrabinska S, Persaud K, Wu Y, Pytowski B, Skobe M (2006) Inhibition of VEGFR-3 activation with the antagonistic antibody more potently suppresses lymph node and distant metastases than inactivation of VEGFR-2. Cancer Res 66:2650–2657

Rockson SG (2012) Update on the biology and treatment of lymphedema. Curr Treat Options Cardiovasc Med 14:184–192

Ruocco V, Schwartz RA, Ruocco E (2002) Lymphedema: an immunologically vulnerable site for development of neoplasms. J Am Acad Dermatol 47:124–127

Saaristo A, Tammela T, Timonen J, Yla-Herttuala S, Tukiainen E, Asko-Seljavaara S, Alitalo K (2004) Vascular endothelial growth factor-C gene therapy restores lymphatic flow across incision wounds. FASEB J 18:1707–1709

Sabin F (1902) On the origin of the lymphatic system from the veins, and the development of lymph hearts and thoracic duct in the pig. Am J Anat 1:367–389

Sabin F (1904) On the development of the superficial lymphatics in the skin of the pig. Am J Anat 3:183–195

Sabin FR (1916) The method of growth of the lymphatic system. Science 44:145–158

Saharinen P, Kerkela K, Ekman N, Marron M, Brindle N, Lee GM, Augustin H, Koh GY, Alitalo K (2005) Multiple angiopoietin recombinant proteins activate the Tie1 receptor tyrosine kinase and promote its interaction with Tie2. J Cell Biol 169:239–243

Schacht V, Ramirez MI, Hong YK, Hirakawa S, Feng D, Harvey N, Williams M, Dvorak AM, Dvorak HF, Oliver G, Detmar M (2003) T1alpha/podoplanin deficiency disrupts normal lymphatic vasculature formation and causes lymphedema. EMBO J 22:3546–3556

Schmid-Schonbein GW (1990) Mechanisms causing initial lymphatics to expand and compress to promote lymph flow. Arch Histol Cytol 53(suppl):107–114

Schoppmann SF, Birner P, Stockl J, Kalt R, Ullrich R, Caucig C, Kriehuber E, Nagy K, Alitalo K, Kerjaschki D (2002) Tumor-associated macrophages express lymphatic endothelial growth factors and are related to peritumoral lymphangiogenesis. Am J Pathol 161:947–956

Schulte-Merker S, Sabine A, Petrova TV (2011) Lymphatic vascular morphogenesis in development, physiology, and disease. J Cell Biol 193:607–618

Shalaby F, Rossant J, Yamaguchi TP, Gertsenstein M, Wu XF, Breitman ML, Schuh AC (1995) Failure of blood-island formation and vasculogenesis in Flk-1-deficient mice. Nature 376:62–66

Shields JD, Emmett MS, Dunn DB, Joory KD, Sage LM, Rigby H, Mortimer PS, Orlando A, Levick JR, Bates DO (2007a) Chemokine-mediated migration of melanoma cells towards lymphatics–a mechanism contributing to metastasis. Oncogene 26:2997–3005

Shields JD, Fleury ME, Yong C, Tomei AA, Randolph GJ, Swartz MA (2007b) Autologous chemotaxis as a mechanism of tumor cell homing to lymphatics via interstitial flow and autocrine CCR7 signaling. Cancer Cell 11:526–538

Shin JW, Min M, Larrieu-Lahargue F, Canron X, Kunstfeld R, Nguyen L, Henderson JE, Bikfalvi A, Detmar M, Hong YK (2006) Prox1 promotes lineage-specific expression of fibroblast growth factor (FGF) receptor-3 in lymphatic endothelium: a role for FGF signaling in lymphangiogenesis. Mol Biol Cell 17:576–584

Skobe M, Hawighorst T, Jackson DG, Prevo R, Janes L, Velasco P, Riccardi L, Alitalo K, Claffey K, Detmar M (2001) Induction of tumor lymphangiogenesis by VEGF-C promotes breast cancer metastasis. Nat Med 7:192–198

Srinivasan RS, Dillard ME, Lagutin OV, Lin FJ, Tsai S, Tsai MJ, Samokhvalov IM, Oliver G (2007) Lineage tracing demonstrates the venous origin of the mammalian lymphatic vasculature. Genes Dev 21:2422–2432

Srinivasan RS, Geng X, Yang Y, Wang Y, Mukatira S, Studer M, Porto MP, Lagutin O, Oliver G (2010) The nuclear hormone receptor Coup-TFII is required for the initiation and early maintenance of Prox1 expression in lymphatic endothelial cells. Genes Dev 24:696–707

Stacker SA, Stenvers K, Caesar C, Vitali A, Domagala T, Nice E, Roufail S, Simpson RJ, Moritz R, Karpanen T, Alitalo K, Achen MG (1999) Biosynthesis of vascular endothelial growth factor-D involves proteolytic processing which generates non-covalent homodimers. J Biol Chem 274:32127–32136

Stacker SA, Caesar C, Baldwin ME, Thornton GE, Williams RA, Prevo R, Jackson DG, Nishikawa S, Kubo H, Achen MG (2001) VEGF-D promotes the metastatic spread of tumor cells via the lymphatics. Nat Med 7:186–191

Stanton AW, Modi S, Mellor RH, Levick JR, Mortimer PS (2009) Recent advances in breast cancer-related lymphedema of the arm: lymphatic pump failure and predisposing factors. Lymphat Res Biol 7:29–45

Stewart FW, Treves N (1948) Lymphangiosarcoma in postmastectomy lymphedema: A report of six cases in elephantiasis chirurgica. Cancer 1:64–81

Szuba A, Skobe M, Karkkainen MJ, Shin WS, Beynet DP, Rockson NB, Dakhil N, Spilman S, Goris ML, Strauss HW, Quertermous T, Alitalo K, Rockson SG (2002) Therapeutic lymphangiogenesis with human recombinant VEGF-C. FASEB J 16:1985–1987

Tammela T, Alitalo K (2010) Lymphangiogenesis: molecular mechanisms and future promise. Cell 140:460–476

Tammela T, Enholm B, Alitalo K, Paavonen K (2005a) The biology of vascular endothelial growth factors. Cardiovasc Res 65:550–563

Tammela T, Saaristo A, Lohela M, Morisada T, Tornberg J, Norrmen C, Oike Y, Pajusola K, Thurston G, Suda T, Yla-Herttuala S, Alitalo K (2005b) Angiopoietin-1 promotes lymphatic sprouting and hyperplasia. Blood 105:4642–4648

Tammela T, Saaristo A, Holopainen T, Lyytikka J, Kotronen A, Pitkonen M, Abo-Ramadan U, Yla-Herttuala S, Petrova TV, Alitalo K (2007) Therapeutic differentiation and maturation of lymphatic vessels after lymph node dissection and transplantation. Nat Med 13:1458–1466

Taniguchi K, Kohno R, Ayada T, Kato R, Ichiyama K, Morisada T, Oike Y, Yonemitsu Y, Maehara Y, Yoshimura A (2007) Spreds are essential for embryonic lymphangiogenesis by regulating vascular endothelial growth factor receptor 3 signaling. Mol Cell Biol 27:4541–4550

Taylor AE, Gibson WH, Granger HJ, Guyton AC (1973) The interaction between intracapillary and tissue forces in the overall regulation of interstitial fluid volume. Lymphology 6:192–208

Tobler NE, Detmar M (2006) Tumor and lymph node lymphangiogenesis—impact on cancer metastasis. J Leukoc Biol 80:691–696

Traboulsi EI, Al-Khayer K, Matsumoto M, Kimak MA, Crowe S, Wilson SE, Finegold DN, Ferrell RE, Meisler DM (2002) Lymphedema-distichiasis syndrome and FOXC2 gene mutation. Am J Ophthalmol 134:592–596

Tso P (1994) Intestinal lipid absorption. In: Johnson L (ed) Physiology of the gastrointestinal tract. Raven Press, New York, pp 1867–1908

Tvorogov D, Anisimov A, Zheng W, Leppanen VM, Tammela T, Laurinavicius S, Holnthoner W, Helotera H, Holopainen T, Jeltsch M, Kalkkinen N, Lankinen H, Ojala PM, Alitalo K (2010) Effective suppression of vascular network formation by combination of antibodies blocking VEGFR ligand binding and receptor dimerization. Cancer Cell 18:630–640

Uhrin P, Zaujec J, Breuss JM, Olcaydu D, Chrenek P, Stockinger H, Fuertbauer E, Moser M, Haiko P, Fassler R, Alitalo K, Binder BR, Kerjaschki D (2010) Novel function for blood platelets and podoplanin in developmental separation of blood and lymphatic circulation. Blood 115:3997–4005

Uldrick TS, Whitby D (2011) Update on KSHV epidemiology, Kaposi Sarcoma pathogenesis, and treatment of Kaposi Sarcoma. Cancer Lett 305:150–162

Veikkola T, Jussila L, Makinen T, Karpanen T, Jeltsch M, Petrova TV, Kubo H, Thurston G, McDonald DM, Achen MG, Stacker SA, Alitalo K (2001) Signalling via vascular endothelial growth factor receptor-3 is sufficient for lymphangiogenesis in transgenic mice. EMBO J 20:1223–1231

von der Weid PY, Rehal S, Ferraz JG (2011) Role of the lymphatic system in the pathogenesis of Crohn's disease. Curr Opin Gastroenterol 27:335–341

Wang HU, Chen ZF, Anderson DJ (1998) Molecular distinction and angiogenic interaction between embryonic arteries and veins revealed by ephrin-B2 and its receptor Eph-B4. Cell 93:741–753

Wang HW, Trotter MW, Lagos D, Bourboulia D, Henderson S, Makinen T, Elliman S, Flanagan AM, Alitalo K, Boshoff C (2004) Kaposi sarcoma herpesvirus-induced cellular reprogramming contributes to the lymphatic endothelial gene expression in Kaposi sarcoma. Nat Genet 36:687–693

Warren AG, Brorson H, Borud LJ, Slavin SA (2007) Lymphedema: a comprehensive review. Ann Plast Surg 59:464–472

Whitehurst B, Eversgerd C, Flister M, Bivens CM, Pickett B, Zawieja DC, Ran S (2006) Molecular profile and proliferative responses of rat lymphatic endothelial cells in culture. Lymphat Res Biol 4:119–142

Wiegand S, Eivazi B, Barth PJ, von Rautenfeld DB, Folz BJ, Mandic R, Werner JA (2008) Pathogenesis of lymphangiomas. Virchows Arch 453:1–8

Wigle JT, Oliver G (1999) Prox1 function is required for the development of the murine lymphatic system. Cell 98:769–778

Wigle JT, Harvey N, Detmar M, Lagutina I, Grosveld G, Gunn MD, Jackson DG, Oliver G (2002) An essential role for Prox1 in the induction of the lymphatic endothelial cell phenotype. EMBO J 21:1505–1513

Wijchers PJ, Burbach JP, Smidt MP (2006) In control of biology: of mice, men and Foxes. Biochem J 397:233–246

Wiley HE, Gonzalez EB, Maki W, Wu MT, Hwang ST (2001) Expression of CC chemokine receptor-7 and regional lymph node metastasis of B16 murine melanoma. J Natl Cancer Inst 93:1638–1643

Wu M, Han L, Shi Y, Xu G, Wei J, You L, Chen Y, Zhu T, Li Q, Li S, Meng L, Lu Y, Zhou J, Wang S, Ma D (2010) Development and characterization of a novel method for the analysis of gene expression patterns in lymphatic endothelial cells derived from primary breast tissues. J Cancer Res Clin Oncol 136:863–872

Xu Y, Yuan L, Mak J, Pardanaud L, Caunt M, Kasman I, Larrivee B, Del Toro R, Suchting S, Medvinsky A, Silva J, Yang J, Thomas JL, Koch AW, Alitalo K, Eichmann A, Bagri A (2010) Neuropilin-2 mediates VEGF-C-induced lymphatic sprouting together with VEGFR3. J Cell Biol 188:115–130

Yamaguchi T, Ichise T, Iwata O, Hori A, Adachi T, Nakamura M, Yoshida N, Ichise H (2008) Development of a new method for isolation and long-term culture of organ-specific blood vascular and lymphatic endothelial cells of the mouse. FEBS J 275:1988–1998

Yamazaki T, Yoshimatsu Y, Morishita Y, Miyazono K, Watabe T (2009) COUP-TFII regulates the functions of Prox1 in lymphatic endothelial cells through direct interaction. Genes Cells 14:425–434

Yaniv K, Isogai S, Castranova D, Dye L, Hitomi J, Weinstein BM (2006) Live imaging of lymphatic development in the zebrafish. Nat Med 12:711–716

Yin N, Zhang N, Xu J, Shi Q, Ding Y, Bromberg JS (2011) Targeting lymphangiogenesis after islet transplantation prolongs islet allograft survival. Transplantation 92:25–30

Yoffey JM, Courtice FC (eds) (1970) The formation of lymph in lymphatics, lymph and the lymphomyeloid complex. Academic, London, pp 123–132

Yoon YS, Murayama T, Gravereaux E, Tkebuchava T, Silver M, Curry C, Wecker A, Kirchmair R, Hu CS, Kearney M, Ashare A, Jackson DG, Kubo H, Isner JM, Losordo DW (2003) VEGF-C gene therapy augments postnatal lymphangiogenesis and ameliorates secondary lymphedema. J Clin Invest 111:717–725

Yuan L, Moyon D, Pardanaud L, Breant C, Karkkainen MJ, Alitalo K, Eichmann A (2002) Abnormal lymphatic vessel development in neuropilin 2 mutant mice. Development 129:4797–4806

Zumsteg A, Baeriswyl V, Imaizumi N, Schwendener R, Ruegg C, Christofori G (2009) Myeloid cells contribute to tumor lymphangiogenesis. PLoS One 4:e7067

Chapter 3
Microvascular Fluid Filtration and Lymph Formation

C.C. Michel

Abstract Pre-nodal lymph is the effluent of tissue interstitial fluid (ISF) and its flow equals the net rate of ISF formation. In most tissues, ISF is formed as an ultrafiltrate of plasma in the capillaries and venules of the microcirculation but in some tissues its composition is dominated periodically (intestinal mucosa) or continuously (kidney) by the absorbate from adjacent epithelia. As Starling argued in 1896, microvascular fluid filtration is determined by a balance of hydrostatic and osmotic pressures across pathways through microvascular endothelium and experimental evidence supporting this concept is reviewed. More recently, it has been appreciated that since ISF is formed by microvascular filtration, steady state osmotic pressure differences across the endothelial pathways are themselves dependent on the filtration rate through them. This ensures that lymph formation continues even when microvascular pressures are low. Further, because the ultrafiltration properties of endothelial pathways for macromolecules differ from those of the main pathways for water and small solutes, the overall rate of ultrafiltration and lymph formation is determined by osmotic pressures across the ultra-filter of the water pathways and this is thought to be the endothelial glycocalyx.

3.1 Introduction

The interstitial fluid (ISF) is continuously renewed at rates that vary from tissue to tissue. Its source, to a greater or lesser extent in all tissues, is the plasma of the microcirculation and it drains into the lymphatics. Although the ISF of the kidneys

C.C. Michel (✉)
Department of Bioengineering, Imperial College, London, SW7 2AZ, UK
e-mail: c.c.michel@imperial.ac.uk

L. Santambrogio (ed.), *Immunology of the Lymphatic System*,
DOI 10.1007/978-1-4614-3235-7_3,

and the intestinal mucosae arise largely as the absorbate of the adjacent epithelia, ISF is formed in most tissues by ultrafiltration of the blood plasma circulating through the capillaries and venules. Ludwig and Noll first proposed that lymph was formed as an ultrafiltrate of plasma but the failure of peripheral lymph flow to correlate with arterial blood pressure led this hypothesis to be superseded by the proposal that lymph was formed as a secretion of the capillary endothelia. In a series of papers published between 1892 and 1896, Ernest Starling showed that when the experiments used to demonstrate the secretion theory were extended, their results provided strong evidence for capillary ultrafiltration as the source of the lymph (Barcroft 1976).

There were also observations suggesting that fluid in tissues could be taken up into the blood. After severe hemorrhage a relatively rapid fall of hematocrit indicated that fluid had passed from the tissue into the vascular system expanding the plasma volume. Starling (1896) went on to show that isotonic sodium chloride solution injected into the tissues of a dog's hind leg could be absorbed into the blood circulating through the limb. When plasma was substituted for salt solution it was not absorbed. Starling concluded that the absorption of fluid from the tissues was driven by the osmotic pressure exerted by the plasma proteins. Making the first measurements of the osmotic pressure of a solution of macromolecules, Starling found that the plasma proteins exerted an osmotic pressure that lay in the range estimated for capillary pressures. The osmotic pressure of a protein solution became known as the colloid osmotic pressure. From these observations Starling (1896) proposed that fluid movements across the walls of microvessel were determined by differences between the hydrostatic pressures and colloid osmotic pressures in the capillary plasma and in the ISF on the other side of the capillary walls.

3.2 Further Experimental Evidence for Starling's Proposal

Starling's hypothesis slowly came to be accepted as the principle mechanism responsible for net fluid movements between the plasma and the ISF. Then in the mid-1920s, a medical student at the University of Pennsylvania, EM Landis, made the first direct measurements of the capillary pressure, P_C, by cannulating single capillaries in the mesentery of a frog with a glass micropipette which was connected to an adjustable manometer. He also devised an ingenious method for estimating the rate of fluid filtration per unit area of capillary wall (J_V/A) in an occluded microvessel (Landis 1927). Landis measured J_V/A and P_C on the same capillary and repeating the measurements on 72 other microvessels in the mesenteries of 11 frogs, he found a strongly positive linear correlation between these variables. Further, he reported that in many vessels, fluid filtration rates were zero when the capillary pressures lay between 7 and 14 cmH_2O, a range which encompassed values previously determined for the colloid osmotic pressure of plasma from the same species of frog. Landis (1930) went on to measure capillary pressures in the nail fold skin of human subjects at heart level and in the mesenteries of rats and guinea pigs. Here he noted the position of the vessel within the capillary bed. The mean values of pressures in vessels at the arterial end of the capillary bed exceeded mean values of plasma colloid osmotic pressure whereas mean values at the venous end were less plasma colloid osmotic pressure. These observations appeared to support the notion that Starling had first articulated that fluid exchange between the plasma and the ISFs could be thought of in terms of fluid filtering from the arterial half of the capillary beds where capillary pressure exceeded plasma colloid osmotic pressure and being absorbed at the venous end where plasma colloid osmotic pressure exceeded the microvascular pressure. The idea was soon adopted by textbook writers who portrayed blood-tissue fluid exchange as a circulation of ISF through the tissues between the arterial and venous ends of tissue microcirculations. A small excess of filtration over absorption was said to be responsible for the flow of lymph from the tissues. Landis (1934) attempted to moderate this concept where he drew attention to the great variations in capillary (microvascular) pressure that occur throughout the body, particularly in animals as large as humans. As we shall see, this popular concept of fluid filtration from the arterial side of the microcirculation and fluid uptake from the tissues on the venous side rarely if ever occurs in reality.

3.3 The Relation Between Capillary Pressure and the Arterial and Venous Pressures

The problem with Ludwig's filtration theory in many peoples' minds was that increases in arterial blood pressure did not increase lymph flow. The weakness of this argument was recognized in the nineteenth century by the experimental pathologist, Julius Cohnheim, who pointed out that the major resistance to blood flow in systemic circulation lay between the arteries and the capillaries (Cohnheim 1882). This meant that a rise of arterial pressure would be accompanied by a rise in capillary pressure only if the resistance to flow between the arteries and capillaries did not increase and the resistance between the capillaries and the veins did not decrease. By contrast he found that an increase in lymph flow always accompanied a rise in venous pressure and concluded that lymph is generated as an exudation of plasma. It was also clearly understood by Starling and Bayliss from experiments they carried out that led to Starling to formulate his hypothesis. Landis's direct measurements of capillary pressure provided clear evidence for the gradients of pressure in systemic vascular beds. Nearly 20 years later, Pappenheimer and Soto-Rivera (1948) related the mean pressures in arteries, capillaries, and veins to the pre- and post-capillary resistances in an equation, which they (and later others) used to estimate mean microvascular pressures in tissues and organ. Their argument ran as follows. If the rate of filtration of fluid from the capillaries is very small compared with the blood flow through them, then flow rate into the microvascular bed is equal to the outflow from it. If P_a, P_c, and P_v are the mean pressures in the arteries, microvessels, and veins of a tissue, F_{in} and F_{out} are the blood flows into and out and r_{a} and r_{V} are the hydraulic resistances between the arteries and the microvessels (pre-capillary) and the microvessels and veins (post-capillary) respectively, then:

$$F_{\text{in}} = \frac{P_{\text{a}} - P_{\text{C}}}{r_{\text{a}}} = F_{\text{out}} = \frac{P_{\text{C}} - P_{\text{V}}}{r_{\text{V}}}$$

Rearranging this expression:

$$P_{\text{C}} = \frac{P_{\text{a}} + P_{\text{V}} \cdot \left(\frac{r_{\text{a}}}{r_{\text{V}}}\right)}{1 + \frac{r_{\text{a}}}{r_{\text{v}}}} \tag{3.1}$$

Equation (3.1) shows that the greater the ratio of pre- to post-capillary resistance, the closer will P_{C} approximate to P_{v} and the less it will be influenced by changes in P_{a}. At this point it is worth saying that in humans and other large mammals, P_{a} and P_{V} vary considerably with the vertical position of the vessel with respect to heart. Below the heart, P_{a} and P_{V} rise linearly with the increased hydrostatic load. Above the heart P_{a} falls similarly but the larger veins collapse, their flow resistance increases and pressure in the venules and smaller veins remains almost constant. Consequently in human subjects, P_{C} of different tissues varies considerably with vertical position with respect to the heart and it has been shown that P_{C} in the feet of human subjects may exceed 100 mmHg when they are standing still (Levick and Michel 1978; Mahy et al. 1995). This simple inference is itself reason for doubting that fluid balance between the blood and the tissues is achieved by filtration from the arterial side of microvascular beds and fluid uptake from the tissues in the venous capillaries and venules.

3.4 Quantitative Expressions of Microvascular Fluid Filtration

The experiments of Landis and Pappenheimer and Soto-Rivera are generally regarded to have established the validity of Starling's hypothesis. They showed that fluid filtration per unit area of microvascular wall (J_{V}/A) was directly proportional to the difference between the microvascular hydrostatic pressure and the plasma colloid osmotic pressures (Π_{C}) and could be described by an equation of the form:

$$\frac{J_{\text{V}}}{A} \approx k(P_{\text{C}} - \Pi_{\text{C}})$$

The constant k was related to the conductance of the channels that carried fluid through the microvascular walls and the number of these channels per unit area.

Pappenheimer and Soto-Rivera (1948) found their data was more accurately described if a small constant pressure of 1–2 mmHg was subtracted from the term $(P_C - \Pi_C)$. They suggested that this was the result of very low concentrations of plasma proteins in the ISF giving rise to a small colloid osmotic pressure of Π_i. They also discussed the possibility that the hydrostatic pressure in the ISF, P_i, might differ from zero so that trans-capillary fluid exchange should be thought of in terms of the differences in hydrostatic and colloid osmotic pressures across the walls of the exchange vessels, i.e., ΔP and $\Delta\Pi$. Thus:

$$\frac{J_V}{A} = k[(P_C - P_i) - (\Pi_C - \Pi_i)] = k(\Delta P - \Delta\Pi) \tag{3.2}$$

During the 1950s there were two very important developments in the theory of fluid exchange across membranes. The first was the description of the permeability properties of microvascular walls to water and small water soluble solutes in terms of flow and diffusion through a population of cylindrical pores with radii having similar values to those of plasma macromolecules (Pappenheimer et al. 1951). Although the pathways for water and hydrophilic solutes are not cylindrical pores, the model was shown to be remarkably robust and the principles involved have been developed and applied to filtration and diffusion through water filled channels of molecular dimensions but with different geometries from cylindrical pores.

The second theoretical advance arose from the application of nonsteady state thermodynamics to the theory of osmotic pressure across membranes, which were permeable to the solute responsible for the osmotic pressure difference (Staverman 1951; Kedem and Katchalsky 1958). It was shown that (3.2) would be valid only if capillary walls were completely impermeable to plasma proteins. When the solute responsible for the osmotic pressure can penetrate the membrane, the osmotic pressure difference across the membrane is reduced by a constant, the reflection coefficient, which is a measure of the membrane's leakiness to the solute. The reflection coefficient is a measure of the "reflected fraction" of the solute molecules when their solution is ultrafiltered through the membrane. Thus, when the membrane is completely impermeable to the solute, the reflection coefficient, which is usually symbolized as σ, has a value of 1 and when a solution of the solute can flow through the membrane without change in its concentration, $\sigma = 0$. From this, (3.2) should be modified slightly to:

$$\frac{J_V}{A} = L_P(\Delta P - \sigma\Delta\Pi) \tag{3.3}$$

L_P is standard symbol for the hydraulic conductivity or hydraulic permeability of a membrane and is now used instead of k.

When considering the exchange of water between the plasma and the ISF, many solutes are present and potentially each solute could contribute to osmotic pressure differences through the product of its reflection coefficient at the microvascular walls and its concentration difference. If there are n solutes in the plasma, fluid exchange is described the expression:

$$\frac{J_{\mathrm{V}}}{A} = L_{\mathrm{P}}\left(\Delta P - \sum \sigma_n \Delta \Pi_n\right) \tag{3.4}$$

In most circumstances the effective osmotic pressure difference is determined by the plasma macromolecules because they are usually present in different concentrations in plasma and ISF and their reflection coefficients are close to unity. For most plasma solutes, σ has a value of 0.1 or less and their concentration differences between across microvascular walls are negligible. Because the terms within the brackets on the right hand side of (3.4) are determined by P_{C}, P_{i}, Π_{C}, and Π_{i} and these hydrostatic and colloid osmotic pressures are called the Starling pressures.

3.5 Measurement of the Extravascular Pressures

Before 1963, it was assumed that interstitial hydrostatic pressure, P_{i}, was close to atmospheric pressure and consequently was of minor importance in determining fluid exchange. It was speculated that after prolonged filtration into tissues P_{i} might increase and attenuate filtration rate. Similarly, although it was known that lymph concentrations of plasma proteins might be 50 % of their plasma concentration, it was assumed that Π_{i} played a minor role in normal fluid exchange. Their relatively high concentrations in lymph was believed to reflect the concentration of the ISF as a result of the absorption of fluid at the venous end of the microcirculation—or even the absorption of fluid from the lymph as it flow through the lymph nodes.

Attempts to measure P_{i} by inserting a fine needle into the tissues were complicated by blockage around the tip of needle preventing free movement of fluid between the tissue and measuring system and hence the uncertainty of determining a balancing pressure. In 1963, Guyton (1963) circumvented this problem by measuring the pressure of fluid filling a hollow rigid capsule that had been implanted and allowed to heal into place subcutaneously. The values he reported for P_{i} were subatmospheric, in the range of −3 to −7 mmHg (Guyton et al. 1971). The wick method, introduced by Scholander et al. (1968) and developed in mammalian physiology by Aukland et al. (1981), also gave subatmospheric pressures for subcutaneous tissues in the range of −1 to −4 mmHg. Both methods showed that P_{i} rose sharply to atmospheric pressure when ISF volume is expanded. Further increase in ISF volume led to little increase in P_{I} until the tissues were clearly edematous. Both wick and capsule methods were used to obtain ISF for estimation of Π_{i}. Values for Π_{i} obtained by these methods approximated closely to those obtained in local pre-nodal lymph when this could be sampled indicating that the relatively high values of Π_{I} were representative of its local average value in the tissues and could no longer be neglected.

3.6 Most Venous Microvessels Appear to Filter Fluid Not Absorb It

Reviewing the subject of capillary filtration, Levick (1991) assembled 14 sets of data where P_{ven}, the pressure in the small veins or post-capillary venules had been measured in addition to P_i, Π_i, and Π_C. For each of these data sets, he calculated P(net), the net pressure driving fluid filtration from the venous end of microcirculation into the tissues as:

$$P(\text{net}) = P_{ven} + \Pi_i - \Pi_C - P_i \quad (3.5)$$

According to the classical view, fluid is absorbed from the tissues in the venous microcirculation so the net pressure driving filtration should be negative. Negative values of P(net) (a pressure favoring fluid uptake from the tissues) were only found in peritubular capillaries of the renal cortex and in capillaries of the intestinal mucosa. In most tissues where values were available, P(net) was found to be positive. These tissues included skeletal muscle, skin and subcutis, lung and various joints. Skeletal muscle constitutes 40 % of the body mass of most mammals and lung has the lowest values for the intravascular pressures. For the two studies on skeletal muscle, P(net) was 3 and 6 mmHg and for lung it was 6 mmHg. Previously, investigators had noted that direct measurements of P_C indicated that fluid filtration was occurring in the vessels in which they measured but it seemed possible that the vessels were atypical of the microcirculation as a whole. By using measurements of venous pressure, Levick assembled persuasive evidence against the textbook picture of the absorption of fluid into venous end of the microcirculation. There were also by this time other arguments against fluid uptake being more than a transient phenomenon in most microvascular beds and these arose from measurements of fluid filtration rates in single capillaries.

3.7 Fluid Exchange Under Transient and Steady State Conditions

Using a development of the Landis red cell method in single capillaries perfused in situ, it is possible to estimate transient rates of fluid filtration following step changes in P_C. When plasma proteins and red cells are present in the perfusate, the relations between J_V/A and P_C reproduce the classical picture of a linear relation with fluid filtration into the tissues with step increases in P_C above Π_C and fluid uptake from the tissues following step decreases in P_C below Π_C. If P_C is held above Π_C, reproducible values for the filtration rate are measured. When P_C is held constant at a value less than Π_C, however, the rates of fluid absorption diminish with time and after several minutes no fluid uptake can be measured (Michel 1980). This asymmetrical behavior of fluid exchange is a consequence of the finite permeability of microvascular walls to the plasma proteins that are responsible for the colloidal osmotic pressures (Michel 1984). It has been known for over 120 years that increasing the capillary filtration rate by raising the local venous pressure increases the flow of lymph and decreases its plasma protein concentrations. The concentration of a plasma macromolecule in the capillary filtrate, C_i, is determined by the ratio of its rate of transport through the endothelial barrier (J_S) to the fluid filtration rate (J_V), i.e.,

$$C_i = \frac{J_S}{J_V} \tag{3.6}$$

Since Π_i is directly proportional to C_i, (3.6) provides an explanation of the observed diminution in fluid uptake with time. If initially P_C is high and C_i and Π_i are low, a step reduction of P_C to values below $\sigma\Delta\Pi$ should mean that fluid uptake occurs. If however, P_C remains below $\sigma\Delta\Pi$ and fluid uptake continues C_i (and hence Π_i) starts to rise as $J_V<0$ (3.6). An increase in Π_i means a decrease in $\Delta\Pi$ and a reduction in fluid uptake from the tissues. This continues until fluid uptake becomes zero and reaches a low level of fluid filtration, just great enough to hold Π_i constant. This is the steady state level of J_V for that value of P_C and Π_C. To calculate the steady value of Π_i, the convection–diffusion equation for J_S was solved for a given value of J_V/A and then divided by J_V to give a steady state value of C_i. This is usually expressed as:

$$C_i = C_P \frac{(1-\sigma)}{(1-\sigma e^{-Pe})} \tag{3.7}$$

Pe, the Peclet number, can be thought as the ratio of the velocity of the solute that is carried through the microvascular wall by fluid filtration to its diffusion velocity through the same channels in the vessel wall. Its importance may be appreciated by remembering that in the absence of fluid filtration, diffusion ultimately leads to C_i increasing until it equals C_p. If the membrane acts as an imperfect molecular sieve during filtration (σ having a value between 0 and 1.0), filtration dilutes the fluid leaving the downstream side of the membrane. If velocity of the solute molecules washed by fluid filtration is slow relative to their velocity of diffusion, the degree of dilution of the solution emerging downstream is minimal. This happens when small ions and small water soluble molecules exchange between the plasma and the ISF for their rates of exchange appear to be independent of fluid filtration. As filtration velocity of solute molecules increases relative to their diffusion velocity, so the dilution of the solution emerging downstream becomes more obvious and when filtration velocity is infinitely faster than diffusion velocity, C_i is reduced to a limiting value of $(1-\sigma)C_P$. Using (3.7) and some elementary algebra, one obtains an expression for $\sigma\Delta\Pi$ which can then be substituted into (3.3) to give an equation describing the steady state relations between J_V/A and ΔP:

$$\frac{J_V}{A} = L_P\left[\Delta P - \sigma^2\Pi_C\frac{(1-e^{-Pe})}{(1-\sigma e^{-Pe})}\right] \tag{3.8}$$

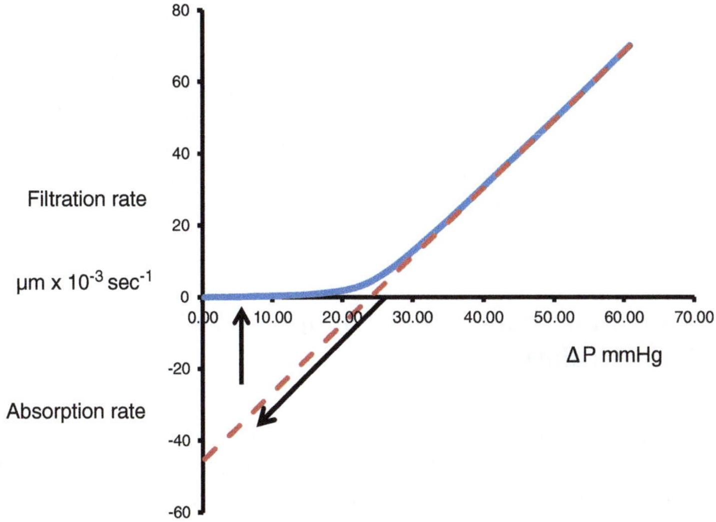

Fig. 3.1 The relations between fluid filtration and the difference in hydrostatic pressure across microvascular walls under transient and steady state conditions. The *dashed line* represents the transient relations and the *solid curve* the steady state. The *arrows* show the changes that follow a step reduction in ΔP from just above the effective osmotic pressure of the plasma (25 mmHg) to well below it: initially there is rapid fluid absorption which attenuates to become a low level of filtration

It is important to note that (3.8) does not allow one to calculate values of J_V/A from values of ΔP because J_V appears in the Pe exponent terms on the right hand side. To see the shape of the relationship, (3.8) is rearranged to give an expression for ΔP in terms of J_V/A (Michel 1984). Figure 3.1 compares the shape of this curve, with the classical linear relation seen under transient conditions. These theoretical arguments were tested in a series of experiment on single mesenteric capillaries perfused with solutions having known values of Π_C (Michel and Phillips 1987). The transient values of J_V/A were first measured at a given P_C. P_C in the vessel was held constant, measurements of J_V/A repeated at regular intervals until they became constant. When initial (transient) values of J_V/A were plotted against P_C the classical linear relation was obtained. When final (steady state) values were plotted, they followed the curvilinear relation predicted by (3.8) with points in the high range of P_C approximating closely to those of the transient relation for that vessel but steady state values of J_V/A were indistinguishable from zero when P_C was in the range that resulted in fluid uptake from the tissues under transient conditions.

A surprising feature of the single capillary experiments was that a steady state was reached quite rapidly (usually <10 min). It has been suggested that this might be a consequence of the confined pericapillary space in mesenteric vessels. It is worth noting that whole organ studies on the lungs show the curvilinear relations between fluid filtration rates and mean capillary pressures that are predicted by (3.8). Since measurements of fluid filtration rates in the whole organ take 5 min here, it suggests that steady state relations of trans-capillary fluid exchange develop rapidly in pulmonary capillaries. Like mesenteric capillaries, the pericapillary spaces of the alveoli are confined, supporting the idea that a small ISF compartment just outside the vessels reaches a steady state more rapidly than the ISF as a whole. In tissues such as skeletal muscle, fluid uptake following a step reduction in P_C diminishes more slowly, approaching a new steady state exponentially with time constants of tens of minutes. It would seem nevertheless that absorption of fluid into the microcirculation is a transient phenomenon in most tissues.

3.8 Steady State Fluid Uptake in Specialized Microcirculations

Steady state fluid uptake does occur, however, in the microcirculations of the renal cortex and medulla and in the intestinal mucosa. There may also be steady state uptake of fluid from the pre-nodal lymph into the microcirculation of the lymph nodes. These tissues all have in an ISF that is rapidly renewed by fluid with a low protein concentration.

As mentioned earlier, most of the ISF of the intestinal mucosa and the kidneys is not formed as a plasma ultrafiltrate but as a protein free epithelial secretion. In the rat ileum, the mucosal microcirculation takes 70 % of the blood flow although P_C of the mucosal capillaries lies well below that of microvessels in the parallel circulation through the smooth muscle. During intestinal absorption, the flow of intestinal lymph increases and its protein concentration falls. Measurements of the Starling pressures (P_C, P_i, Π_C, and Π_i) in mucosal capillaries under these conditions reveal that P_C is considerably less than P(net) indicating fluid uptake into the blood (Gore and Bohlen 1978). Although most of the fluid absorbed by the epithelium is taken up by the capillaries, the accompanying brisk lymph flow ensures that proteins that leak into the ISF from the capillaries are not allowed to accumulate and C_i is kept low (Granger and Taylor 1978). In the non-absorbing ileum, however, Π_i rises and fluid absorption into the mucosal microvessels reverts to low levels of filtration (Granger and Taylor 1978).

In both the cortex and medulla of the kidney, protein free fluid is continuously secreted into the ISF by the epithelia of the proximal and distal tubules and the collecting ducts diluting the plasma protein that enters the ISF from the blood. In the cortex, fluid that is not absorbed by the peritubular capillaries flows away in the lymph which carries the plasma protein with it (Ulfendahl and Wolgast 1992). In the renal medulla, where there are no lymphatics, the relatively high flow of protein free fluid into the ISF from the collecting ducts is sufficient to keep Π_i so low (and P_i high) that not only is fluid continuously absorbed into the ascending vasa recta but at a sufficient rate to carry ISF protein with it (MacPhee and Michel 1995; Pallone 1992; Tenstad et al. 2001).

Fluid is also taken up into the microcirculation of lymph nodes (Adair and Guyton 1983). While far less is known about fluid exchange in the nodal microcirculation, it appears that the ISF is largely formed by the pre-nodal lymph flowing into the nodes, and is drained away as the post-nodal lymph. The flow rate through the node is high enough to prevent the protein concentration from rising above the critical value that would arrest fluid uptake into the nodal microcirculation. When the protein concentration of the pre-nodal lymph is low, up to half its volume can be absorbed into the nodal microcirculation. A more complete account of microvascular fluid uptake is given in the review of Levick and Michel (2010).

3.9 Net Filtration Calculated from Starling Pressures Overestimates Lymph Flow

From what has been said so far, it seems that in most tissues, the steady state values of the Starling pressures favor a low level of fluid filtration from the microvasculature with transient periods of fluid uptake. Since the filtration capacity of the microcirculation in several tissues has been estimated, the question arises as to whether a driving pressure of 3–6 mmHg (the values of *P*(net) in many tissues) is sufficient to account for the formation of the lymph (Levick 1991). First it should be noted that the values of *P*(net) are minimal values, valid only if σ to plasma protein approximates to 1. Making this assumption, it is instructive to calculate the volume of lymph that might be formed by a driving pressure of 3 mmHg in skeletal muscle capillaries over a period of 24 h. The rate of fluid filtration per unit mass of tissue per mmHg or the capillary filtration capacity (CFC) has been determined many times in skeletal muscle of different mammalian species (including humans) and the reflection coefficient to plasma proteins lies in the range 0.95–1.0 for undamaged healthy tissues. The soft tissues of the human forearm are largely skeletal muscle and compared with smaller mammals, CFC here has a low value of 3.5×10^{-3} mL m in^{-1} mmHg^{-1} 100 g^{-1}. With *P*(net) of 3 mmHg one would estimate approximately 0.1 mL of filtrate would be formed in each kilogram of muscle every minute. If skeletal muscle constitutes 40 % of the body mass, a 70 kg human subject should have 28 kg of muscle from which 2.8 mL of fluid should be filtering every minute and just over 4 L of fluid should be formed over 24 h. We do not have values of lymph flow from the skeletal muscle over 24 h but we do have estimates of lymph flow for the entire body. Values of 1–2 L per 24 h have been measured in human patients with thoracic duct fistulae and total flow in healthy human subjects has been estimated to lie between 2 and 4 L per 24 h (Landis and Pappenheimer 1963). From comparative studies on mammals, Yoffey and Courtice (1970) concluded that volume of lymph over a 24 h period was 50–100 % of the plasma volume, the majority of which is derived from liver and the gastro-intestinal tract. Our calculation of capillary filtration in skeletal muscle yields a value comparable with lymph flow for the entire organism. Allowing for the approximations involved in calculations of this kind, it is clear that the two estimates are inconsistent with each other. For reasonable agreement to be achieved, *P*(net) for skeletal muscle should be one tenth of the value that the simple calculation suggests.

Two explanations have been suggested to account for this discrepancy. Levick's initial proposal (1991) was that in tissues where there is arterial vasomotion, filtration occurs only during periods of flow through microvessels (when P_C has its measured value) and periods when flow is reduced or zero, P_C falls low enough for transient fluid absorption to occur. This idea was originally suggested by Chambers and Zweifach (1947) and has been developed theoretically by Intaglietta and Endrich (1979). In the context of lymph flow it suggests that the capillary ultrafiltrate in a tissue is formed at any one moment from only a fraction of microvascular exchange vessels. Not all tissues, however, show the clear vasomotor patterns necessary for such a model. Also, it is unlikely that P_C would fall below P_{ven} during the "no-flow phase" of the vasomotor cycle.

The second possible explanation concerns the appropriate value of Π_i that is used to estimate *P*(net). The colloid osmotic pressure difference relevant to the calculation of *P*(net) is that exerted across the molecular filter of the endothelium. The present view is that the glycocalyx at the luminal surface of the endothelial cells is the ultrafilter (Michel and Curry 1999) and we will now consider how the concentration difference of macromolecules across this structure may differ considerably from the difference in mean concentrations between plasma and ISF.

3.10 Osmotic Pressure Differences Across the Endothelial Ultrafilter

The glycocalyx: The glycocalyx on the luminal surfaces of endothelial cells is complex network of glycosaminoglycan chains (syndecan-1, glypican and hyaluronan) and sialoglycoproteins, extending from 60 to 600 nm from the luminal cell membrane. It has been shown that the basic architecture of the glycocalyx of fenestrated endothelia (including that over the fenestrae) is similar to that covering continuous endothelial cells and their intercellular clefts (Squire et al. 2001; Arkill et al. 2011). It has a regular arrangement of molecular fibers with diameters in the range 12 nm set perpendicular to the membrane surface and separated from their neighbors by a distance of 8 nm. There is also some evidence that fibrous chains of similar dimensions lie parallel to the membrane. Such an arrangement should act as a molecular filter having a high reflection coefficient to a molecule such as serum albumin, similar to the microvascular wall itself. This conclusion bolsters the hypothesis that the luminal endothelial glycocalyx is the molecular filter for both fenestrated and continuous endothelia and is consistent with the exclusion of plasma macromolecules from the cell membranes on luminal surfaces of endothelia (Vink and Duling 1996, 2000). If the fiber-matrix theory is correct, the value of $\Delta\Pi$ that determines fluid exchange across microvascular walls is the difference between Π_C and the colloid osmotic pressure on the downstream side of the glycocalyx, Π_G. It was suggested independently by Michel (1997) and Weinbaum (1998) that differences between Π_G and Π_i could account for the discrepancies between estimates of net filtration rates and measurements of lymph flow.

Fenestrated endothelia: In most tissues where the microvessels have fenestrated endothelia, the Starling pressures appear to be consistent with the general picture of fluid exchange. The exceptions are two synovial joints where the pressures have been estimated and P(net) indicates that fluid filtration rates are much greater than joint lymph flow. McDonald and Levick (1993) and Levick (1994) investigated the effects on filtration rate of changing $\Delta\Pi$ by varying Π_C and Π_i independently. They found that changes in $\Delta\Pi$ brought about changes in Π_C were three times greater than changes induced by Π_i with Π_C held constant. Their hypothesis that this was a consequence of gradients of macromolecule concentration on the serosal side of the synovial capillaries was supported by the results of a 2D model of synovial fluid exchange. The latter also predicted several nonintuitive consequences that were consistent with their observations (Levick 1994).

Continuous endothelia: The largest discrepancies between predictions of filtration rates from $[P(\text{net}) - P_V]$ differences and lymph flow have been found in muscle, skin, mesentery, and lung where the microvascular endothelia are continuous. There are three pathways for water through this type of endothelium: 85–90 % of L_P is accounted by the route through the glycocalyx and then through the intercellular clefts; a second route is via the aquaporin channels in the cell membranes which contribute up to 10 % to the total L_P; a third pathway for macromolecules, often referred to as "the large pores," may also act as a fluid conducting route. In the argument that follows, it is assumed that this third path for macromolecules is spatially separated from the main pathways through the intercellular clefts. Obviously since the large pores are a transport route for plasma proteins, macromolecules cannot have been filtered out of the fluid by the glycocalyx before it flows through this pathway.

Once the fluid has filtered through the glycocalyx and before it enters the intercellular cleft, it contains albumin and larger proteins at only 1–2 % of their plasma concentrations. Potentially this means that nearly the full osmotic pressure of the plasma proteins can be exerted across the glycocalyx. The probability of protein, that has entered the ISF by the "large pores" can enter the clefts from the tissue side and diffuse back to reduce this concentration difference, is reduced very considerably by the amplification of fluid velocity through the clefts.

First, the entrances to the clefts at the luminal surface cover a few thousandths of the area of the endothelium. By just being diverted into the clefts fluid velocity is amplified several hundred fold over its mean value across the vessel wall. Flow through the cleft is further restricted by the tight junctions which funnel flow through the breaks in the junctional strands. Since these openings account for no more than 10 % and probably closer to 1 % of the length of junctions, fluid velocity is amplified a further 10–100 times (Adamson and Michel 1993; Adamson et al. 2004). Through the breaks in the junctional strands, fluid velocity is more than a thousand times greater than its mean value per square centimeter of endothelium. Calculations indicate that this large amplification of fluid velocity reduces the rate at which protein in the ISF can back-diffuse through the intercellular clefts and a pressure difference of 1 mmHg is sufficient to maintain a substantial difference in protein concentration between the entrance to clefts below the glycocalyx and the ISF (Michel 1997; Weinbaum 1998). This is shown diagrammatically in (Fig. 3.2).

Evidence for gradients of protein concentration within the intercellular cleft pathway: Weinbaum and his colleagues (Hu and Weinbaum 1999; Adamson et al. 2004)

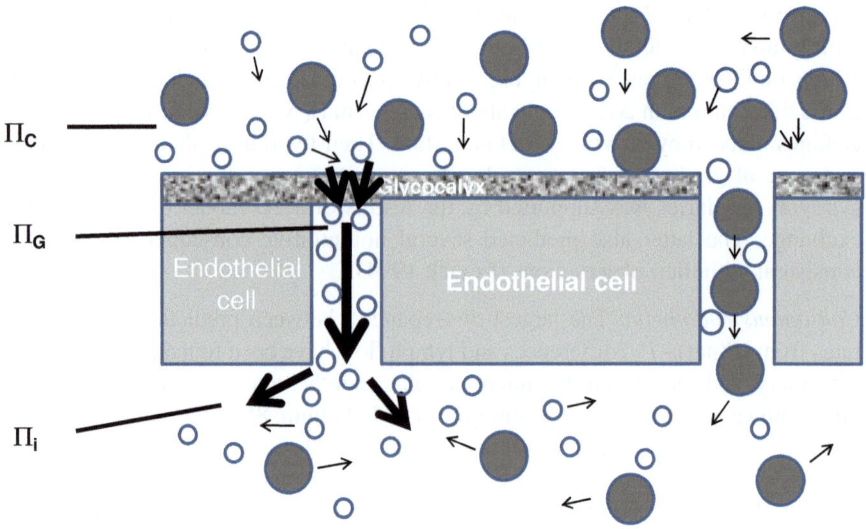

Fig. 3.2 A schematic diagram of fluid and protein exchange across the walls of microvessels with continuous endothelium. *Small open circles* represent water molecules, *large filled circles* protein, with the plasma shown above the endothelium and the ISF below. Water crosses the wall largely via the intercellular cleft (on the *left* of the diagram) after it has passed through the overlying glycocalyx which sieves out the protein; proteins cross via the "large pore" passing through the right hand side of the endothelial cell below an opening in the glycocalyx. The thickness and length of the *arrows* indicate mean velocities of the molecules. Because the velocity of the fluid flowing through the intercellular clefts is amplified by 3–4 orders of magnitude, it prevents protein molecules from diffusing back from the ISF to the underside of the glycocalyx. Consequently the osmotic pressure difference across the glycocalyx is close to that of the plasma and largely independent of that of the ISF

followed up the preliminary calculations of gradients of protein concentration within the intercellular clefts downstream from the glycocalyx with a 3D model which was able to examine the steady state relations and fit the observations and results of the simple model of Michel and Phillips (1987). Some of this theoretical work was developed alongside experiments devised by Curry and Adamson to test the hypothesis (Hu et al. 2000; Adamson et al. 2004). The classical Starling hypothesis predicts that raising the ISF protein concentration to its value inside the capillary should abolish any osmotic pressure difference opposing filtration. The Michel–Weinbaum hypothesis predicts the very presence of filtration should reduce Π_G, on the downstream side of the glycocalyx, to less than Π_i. Testing the predictions on frog mesenteric capillaries perfused in situ with an albumin solution, they estimated fluid filtration rates over a range of pressures first with the ISF protein concentration kept low and then with the same concentration of albumin in the ISF as was present in the solution perfusing the capillary (Hu et al. 2000). In a later study on rat mesenteric venules (Adamson et al. 2004) they not only confirmed their earlier findings but extended them to show the steady state relations were also only slightly changed by the high ISF albumin concentration. Measurements of filtration rates across confluent monolayers of cultured endothelial cells have also shown that the effective colloid osmotic pressure opposing filtration is dominated by the mean concentration of protein upstream (Pang and Tarbell 2003).

3.11 Further Developments

One of the big surprises in the experimental demonstration of the steady state relations between fluid filtration and changes in P_C was the speed with which a steady state was established. A possible explanation for this has arisen from the modeling studies and a closer consideration of the structure of the tissues surrounding mesenteric microvessels. To simplify the changes in scale necessary to model events in the intercellular clefts and in the wider ISF, Zhang et al. (2008) devised a simplified 1D model. This led them to argue that mesenteric capillaries are closely invested with pericytes with a relatively small volume of ISF between. Because this confined region between the intercellular clefts of the microvascular endothelium and with the ISF as a whole, it acts as a small compartment that could equilibrate more quickly with the capillary filtrate than the ISF as a whole. In this way they could account for the relatively rapid approach to a steady state reported by Michel and Phillips (1987) which they confirmed. It seems possible that differences in the rate at which a steady state of fluid exchange is approached in different tissues may be related to the different arrangement of cells surrounding microvessels and consequent differences in the micro-architecture the interstitial space here.

3.12 Conclusions

The classical Starling principle states that fluid filtration rates between plasma and ISF are determined by differences in the hydrostatic and colloid osmotic pressures across microvascular walls. It is now appreciated that the colloid osmotic pressures differences concerned are those across the glycocalyx at the luminal surface of microvascular endothelial cells because this is now recognized as the endothelial barrier to macromolecules. This difference may be greater or less than that calculated from the colloid osmotic pressures of the plasma and its mean value for ISF and the so-called "revised Starling principle" is a redefinition of the pressure differences involved.

Furthermore, there is no evidence for (and considerable evidence against) the twentieth century textbook figure of fluid filtering from the arterial end of microvessels and being absorbed from the ISF at their venous ends. It appears that a low level of fluid filtration occurs throughout the entire microcirculation of most tissues and low filtration is their steady state condition. Whereas continuous fluid uptake into the microcirculation does occur in specialized tissues such as in the kidney where the bulk of the ISF is generated as a protein free secretion of the adjacent epithelial cells, fluid uptake into the microcirculation is a transient phenomenon. Thus regulation of ISF volume in the long term is critically dependent on lymphatic drainage.

References

Adamson RH, Lenz JF, Zhang X, Adamson GN, Weinbaum S, Curry FE (2004) Oncotic pressures opposing filtration across non-fenestrated rat microvessels. J Physiol 557:889–907
Adamson RH, Michel CC (1993) Pathways through the intercellular clefts of frog mesenteric capillaries. (Appendix by Parker KH, Phillips CG, Wang W). J Physiol 466:303–327
Adair TH, Guyton AC (1983) Modification of lymph by lymph nodes. II. Effect of increased lymph node venous blood pressure. Am J Physiol 245:H616–H622
Arkill KP, Knupp C, Michel CC, Neal CR, Qvortrup K, Rostgaard J, Squire JM (2011) Similar endothelial glycocalyx structures in microvessels from a range of mammalian tissues: evidence for a common filtering mechanism. Biophys J 101(5):1046–1056
Aukland K, Fadnes HO, Noddeland H, Reed RK (1981) Edema preventing mechanisms in subcutis and skeletal muscle. In: Hargens AR (ed) Tissue fluid pressure and composition. Williams & Wilkins, Baltimore, MD, pp 87–93
Barcroft H (1976) Bayliss-Starling Memorial Lecture 1976. Lymph formation by secretion of filtration? J Physiol (Lond) 260:1–20
Chambers R, Zweifach BW (1947) Intercellular cement and capillary permeability. Physiol Rev 27:436–463
Cohnheim J (1882) Lectures of general pathology, vol 1, The circulation English translation. The New Sydenham Society, 1889, London
Gore RW, Bohlen HG (1978) Microvascular pressures in rat intestinal muscle and mucosal villi. Am J Physiol 233:H685–H693
Granger DN, Taylor AE (1978) Effects of solute coupled transport on lymph flow and oncotic pressure in the cat ileum. Am J Physiol 235:E429–E436
Guyton AC (1963) A concept of negative interstitial pressure based on pressures in implanted perforated capsules. Circ Res 12:399–414
Guyton AC, Granger HJ, Taylor AE (1971) Interstitial fluid pressure. Physiol Rev 51:527–563
Hu X, Adamson RH, Lui B, Curry FE, Weinbaum S (2000) Starling forces that oppose filtration after tissue oncotic pressure is increased. Am J Physiol 279:H1724–H1736
Hu X, Weinbaum S (1999) A new view of Starling's hypothesis at the microstructural level. Microvasc Res 58:281–304
Intaglietta M, Endrich B (1979) Experimental and quantitative analysis of microcirculatory water exchange. Acta Physiol Scand Suppl 463:59–66
Kedem O, Katchalsky A (1958) Thermodynamic analysis of the permeability of biological membranes to non-electrolytes. Biochim Biophys Acta 27:229–245
Landis EM (1927) Micro-injection studies of capillary permeability. II. The relation between capillary pressure and rate at which fluid passes through the walls of single capillaries. Am J Physiol 82:217–238
Landis EM (1930) Micro-injection studies of capillary blood pressure in human skin. Heart 15:209–228
Landis EM (1934) Capillary pressure and capillary permeability. Physiol Rev 14:404–481
Landis EM, Pappenheimer JR (1963) Exchange of substances through the capillary walls. In: Hamilton WF, Dow P (eds) Handbook of physiology. Section 2, circulation II. American Physiological Society, Washington, DC, pp 961–1034
Levick JR (1991) Capillary filtration-absorption balance reconsidered in the light of extravascular factors. Exp Physiol 76:825–857
Levick JR (1994) An analysis of the interaction between extravascular plasma proteins, interstitial flow and capillary filtration; application to synovium. Microvasc Res 47:90–125
Levick JR, Michel CC (1978) The effects of position and skin temperature on the capillary pressure in the fingers and toes. J Physiol 274:97–109
Levick JR, Michel CC (2010) Microvascular fluid exchange and revised Starling principle. Cardiovasc Res 87:198–210

McDonald JN, Levick JR (1993) Effect of extravascular plasma protein on pressure-flow relations across synovium in anaesthetized rabbits. J Physiol 465:539–559
MacPhee PJ, Michel CC (1995) Fluid uptake from the renal medulla into the ascending vasa recta of anaesthetized rats. J Physiol 487:169–183
Mahy IR, Tooke JR, Shore AC (1995) Capillary pressure during and after incremental venous elevation in man. J Physiol 485:213–219
Michel CC (1980) Filtration coefficients and osmotic reflexion coefficients of the walls of single frog mesenteric capillaries. J Physiol 309:341–355
Michel CC (1984) Fluid movements through capillary walls. In: Renkin EM, Michel CC (eds) Handbook of physiology: section 2, vol 4, Microcirculation. American Physiological Society, Bethesda, MD, pp 375–409
Michel CC (1997) Starling: the formulation of his hypothesis of microvascular fluid exchange and its significance after 100 years. Exp Physiol 82:1–30
Michel CC, Curry FE (1999) Microvascular permeability. Physiol Rev 79:703–761
Michel CC, Phillips ME (1987) Steady state filtration at different capillary pressures in perfused frog mesenteric capillaries. J Physiol 388:421–435
Pallone TJ (1992) Molecular sieving of albumin by the ascending vasa recta wall. J Clin Invest 90:30–34
Pang Z, Tarbell JM (2003) In vitro study of Starling's hypothesis in a cultured mono-layer of bovine aortic endothelial cells. J Vasc Res 40:351–358
Pappenheimer JR, Soto-Rivera A (1948) Effective osmotic pressure of the plasma proteins and other quantities associated with the capillary circulation in the hind limbs of cats and dogs. Am J Physiol 152:471–491
Pappenheimer JR, Renkin EM, Borrero LM (1951) Filtration, diffusion and molecular sieving through peripheral capillary membranes. A contribution to the pore theory of capillary permeability. Am J Physiol 167:13–46
Scholander PF, Hargens AR, Miller SL (1968) Negative pressure in the interstitial fluid of animals. Science 161:321–328
Squire J, Chew M, Nnedji G, Neal C, Barry J, Michel CC (2001) Quasi-periodic substructure in microvessel endothelial glycocalyx: a possible explanation for molecular filtering? J Structural Biol 136:239–255
Starling EH (1896) On the absorption of fluids from connective tissue spaces. J Physiol 19:312–326
Staverman AJ (1951) The theory of measurement of osmotic pressure. Recl Trav Chim Pays Bas 70:344–352
Tenstad O, Heyeraas KJ, Wiig H, Aukland K (2001) Drainage of plasma proteins from the renal medullary interstitium in rats. J Physiol 536:533–539
Ulfendahl HR, Wolgast M (1992) Renal circulation and lymphatics. In: Seldin DW, Giebisch G (eds) The kidney: Physiology and pathophysiology. Raven Press Ltd. New York. pp 1017–1047
Vink H, Duling BR (1996) Identification of distinct luminal domains for macromolecules, erythrocytes and leukocytes within mammalian capillaries. Circ Res 79:581–589
Vink H, Duling BR (2000) Capillary endothelial surface layer selectively reduces plasma solute distribution volume. Am J Physiol Heart Circ Physiol 278:H285–H289
Weinbaum S (1998) Distinguished lecture. Models to solve the mysteries in biomechanics at a cellular level. A new view of fiber matrix layers. Ann Biomed Eng 26:627–643
Yoffey JM, Courtice FC (1970) Lymphatics, lymph and the lymphomyeloid complex, chapter 3. Academic, London, p 166
Zhang X, Adamson RH, Curry FE, Weinbaum S (2008) Transient regulation of transport by pericytes in venular microvessels via trapped microdomains. Proc Natl Acad Sci USA 105:1374–1379

Chapter 4
Cellular Composition of Lymph

Andrew M. Platt and Gwendalyn J. Randolph

Abstract This chapter discusses the cellular composition of lymph beginning with a historical perspective into the evolution of our understanding of lymph and the diversity of cells that traffic within it. The biological roles and importance of immune cell transit through the lymphatic vasculature is reviewed in the context of immunity, but with emphasis on new emerging information delineating its importance for the regulation of the immune system. Furthermore, the chapter highlights new insights into how immune cell traffic through lymph influences the function of the entire lymph node itself. The disequilibrium of lymphocyte subtypes in afferent and efferent lymph is discussed in light of the importance these trafficking patterns have on immunity, immunological memory, and the maintenance of tolerance to self-antigen. Trafficking of dendritic cells through the lymphatic vasculature, the importance of delivering these cells to the correct locale, and the mechanisms governing this process are all discussed in detail.

4.1 Introduction

Lymph nodes are a major component of the secondary lymphoid system and they serve as the integrating unit of the lymphatic vasculature. As first demonstrated in 1970, the afferent (pre-nodal) lymphatic vasculature efficiently supports the mobilization of cells from the parenchyma of various organs to draining lymph nodes (Kelly 1970), where they communicate critical information regarding the immunological status of the upstream microenvironment to the lymph node along with

---

A.M. Platt (✉)
Institute of Infection, Immunity and Inflammation, University of Glasgow, Glasgow, UK
e-mail: Andrew.platt@glasgow.ac.uk

G.J. Randolph
Department of Pathology & Immunology, Washington University in St. Louis, St. Louis, MO 63110, USA

L. Santambrogio (ed.), *Immunology of the Lymphatic System*,
DOI 10.1007/978-1-4614-3235-7_4, © Springer Science+Business Media New York 2013

soluble components of the lymph. While lymph nodes collect cells draining tissues through their afferent lymphatic input, release of immune cells from the lymph node through efferent lymphatic vessels (post-nodal) disperses lymphocytes back into the bloodstream, though often not without passage through additional lymph nodes arranged in chains (Braun et al. 2011).

Primarily through the study of dendritic cell migration to lymph nodes, we now know that cell entry into afferent lymphatic vessels is a tightly controlled, highly selective process as detailed in recent reviews (Forster et al. 2012). In brief, DCs use CCR7 to dock onto lymphatic endothelial cells that express the CCR7 ligand CCL21 (Braun et al. 2011; MartIn-Fontecha et al. 2003; Tal et al. 2011). Then they preferentially seek out areas along lymphatic capillaries that have sparse basement membrane (Pflicke and Sixt 2009; Tal et al. 2011) and gain access to the lymphatic lumen through specialized "button-like" junctions found between lymphatic endothelial cells (Baluk et al. 2007). Migration of lymphocytes into efferent lymph leading away from lymph nodes is also a highly regulated process that involves coordination of signals from CCR7 and sphingosine phosphate receptor 1 (Cyster and Schwab 2012). This summary will not cover these regulatory steps in detail but instead will focus on a discussion of the cellular composition of lymph. Where it seems obvious that a major purpose of cell trafficking through lymphatic vessels is induction and regulation of immune responses, a goal of this chapter will be to provide a fresh perspective on the importance and biological roles of cell trafficking through lymph, highlighting advances and surprises that have come through recent research.

4.2 Foundational Observations

Although we are still learning about the types of cells circulating through lymph and how these change during disease, investigators have been examining the cellular content of lymphatic vessels for decades. With regard to understanding the cellular composition of lymph, the sheep has proved to be an invaluable model that has greatly advanced the field. A study performed in resting sheep dating back to 1970 reported results following cannulation of afferent lymphatic vessels draining a variety of body sites including skin, liver, kidney, ovary, testis, and thyroid and efferent lymphatic vessels draining liver or various skin sites (Smith et al. 1970). To our knowledge, this study remains the most comprehensive analysis of the cellular traffic through afferent lymph with respect to the number of different tissues analyzed. The vast majority of afferent lymph cells from all different sites were lymphocytes, whereas 5–20 % of cells in the afferent lymph were deemed to be macrophages with a peculiar veiled morphology (Smith et al. 1970). With more analysis (Drexhage et al. 1979; Pugh et al. 1983; Spry et al. 1980), it soon became clear that at least the majority of these "macrophages" were in fact better classified as dendritic cells, a cell type first described by Ralph Steinman and Zanvil Cohn in 1973 (Steinman and Cohn 1973) and discussed in more detail below. Besides lymphocytes and dendritic cells, a very small number of other cell types were described in resting afferent lymph, including neutrophils, eosinophils, basophils, and plasma cells that collectively comprise 1–10 % of the resting lymph cellular content from all sites studied (Smith et al. 1970).

With regard to rates of cell flux in cannulated lymph, the seminal study by Smith et al. mentioned above showed that at most body sites, for both afferent and efferent lymph, flow rates were on the order of 1–5 mL/h in the sheep. Flow rates from afferent lymph draining the testis were markedly higher, at 10–30 mL/h, but lower from the thyroid, at less than 1 mL/h (Smith et al. 1970). However, the density of cells in lymph per mL was fairly similar at all body sites, ranging from 100 to 1,000 cells per mL. The sole exception with regard to density of cell per mL of lymph was an order of magnitude greater cell number per mL of lymph in afferent vessels draining the liver. Though these rates of cell flux through lymph might at first seem low, when one considers the total amount of lymph that drains an entire organ like skin or liver, the flux of cells through lymph is impressive. Smith et al. estimated that the sheep liver afferent lymph hosts the flux of two billion cells per day (Smith et al. 1970). Rates of cell flux in efferent lymph were generally an order of magnitude higher than rates of transit through afferent lymph (Mackay et al. 1988; Smith et al. 1970). This observation led Mackay and others to early proposals, borne out through the years to be correct, that the input of lymphocytes into lymph nodes through the high endothelium of the lymph node blood vasculature contributes significantly to the character of cells exiting lymph nodes through the efferent lymph and that endothelial cells of the blood and possibly lymphatic vasculature regulate patterns of trafficking (Mackay et al. 1988).

With an established understanding that the majority of afferent lymph cells entering afferent lymph were lymphocytes, a key subsequent advance in the field came from analysis of the types of lymphocytes recovered from afferent lymph. Afferent lymph T cells of the standard $\alpha\beta$ subtype obtained from resting sheep are enriched in CD4+ T cells over CD8+ T cells by 4–5 times relative to their ratios in blood (Hein et al. 1987; Mackay et al. 1990). T cells of the $\gamma\delta$ type are also enriched in afferent lymph (Mackay et al. 1990). Most importantly, T cells recovered from afferent lymph were of the memory phenotype and many had proliferated within the several days leading up to their collection from lymph, but lymphocytes in efferent lymph were overwhelmingly comprised of naïve T cells. The rate of lymphocyte transit through afferent lymph increases tenfold when the upstream tissue has become chronically inflamed (Brown et al. 2010).

These data led to the concept that afferent lymph T cells are recirculating memory cells that survey tissues and re-enter the afferent lymphatics as part of their survey program. By comparison, efferent lymph is naturally enriched in naïve lymphocytes that recirculate through lymph nodes after entering the nodes through high endothelial venules (HEVs) (Mackay et al. 1990), typically employing L-selectin for such passage (Butcher and Picker 1996).

The role of B lymphocyte trafficking in lymph was downplayed in early studies (Mackay et al. 1990), though B cells were described in human afferent lymph (Olszewski et al. 1995). Revisiting this issue in sheep with newer mAbs available to characterize cell phenotypes, Debes and coworkers recently uncovered evidence for

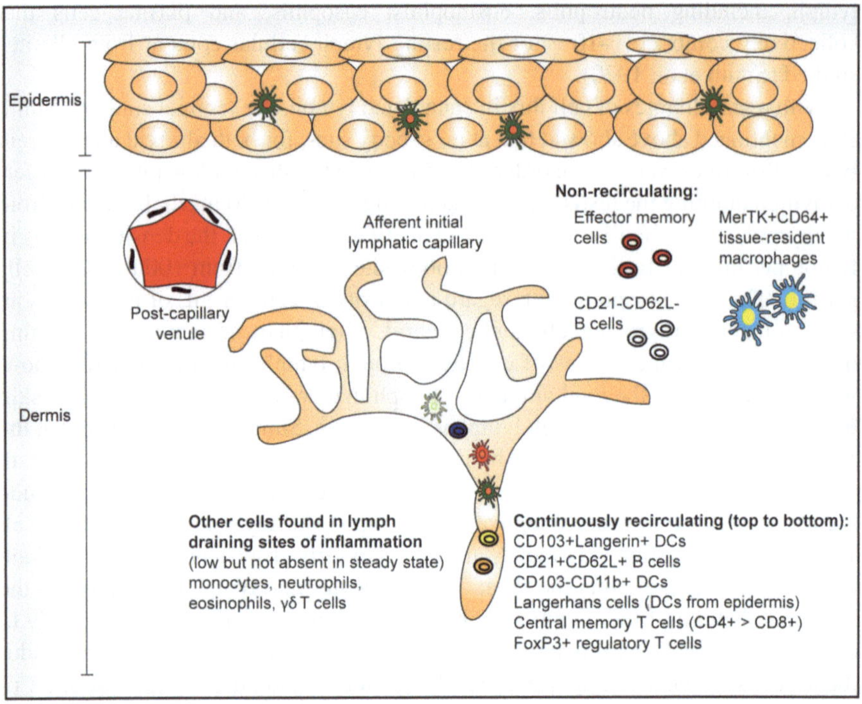

Fig. 4.1 The disequilibrium in immune cell homing through afferent lymph. Schematic depicts a cross-sectional view of skin. Several immune cell types that fail to emigrate out of skin are shown, counterbalanced by a number of immune cell subsets that prominently enter the blind-ended lymphatic capillaries and progress through afferent lymph. A few cell types are quite rare in lymph but increase in inflammation, including granulocytes and monocyte-derived cells, along with a large increase in all the cell types found in lymph under resting conditions

substantial numbers of resting B cells residing in skin and skin-draining lymph (Geherin et al. 2012). Some of these B cells that constitutively entered resting lymph were of the innate B1 type of B lymphocytes (Geherin et al. 2012). In conditions of chronic inflammation, B cell flux into the skin and out of it through afferent lymphatic vessels increased several fold over baseline (Geherin et al. 2012). Thus, the older notion that B cells do not traverse lymph in any great number appears incorrect (Fig. 4.1).

4.3 The Disequilibrium of Lymphocyte Recirculation into Afferent Lymph

Many B cells recovered from afferent lymph express CD62L (L-selectin) and CD21, markers associated with B cells that recirculate between lymph nodes in general. $CD21^-$ and $CD62L^-$ lymphocytes in efferent lymph do not recirculate and have been known to accumulate in spleen. That most resting skin B cells are CD21⁻ and CD62L⁻ raises the possibility that there are two types of B cells that enter resting skin: those with the ability to subsequently leave the skin through lymphatics and therefore seemingly associated with a transient tissue surveillance function and those that take up permanent residence in the skin.

A similar concept appears to apply to T lymphocytes. The entry of memory T lymphocytes into nonlymphoid tissues is not antigen dependent. Lymphocytes recruited to tissues that subsequently engage antigen-presenting cells in antigen-specific interactions are thereby retained, but others exit the tissue through the lymphatics (Jennrich et al. 2012). Many of those engaged by antigen eventually undergo contraction but a small population remains and exists in disequilibrium with recirculating lymphocytes (Gebhardt et al. 2009). Indeed, recent studies indicate that powerful effector memory T cells in the skin of humans and mice do not recirculate (Clark et al. 2012; Jiang et al. 2012). Instead, only central memory T cells exhibit recirculation capacity (Fig. 4.1). Specifically, treatment of humans with anti-CD52 depleted all T cells from the blood and central memory T cells (T_{CM}) from the skin, but skin-resident T effector memory cells (T_{EM}) remained in normal numbers. Anti-CD52 depletion was largely neutrophil dependent, suggesting that T cell depletion was a result of antibody-dependent cytotoxicity, wherein the killer cell was the neutrophil, abundant in blood but rare in skin. Hence, cells bearing CD52 would be depleted only so long as they circulated through blood. By contrast, skin-resident T_{EM} survive because they do not recirculate between blood and tissue (Clark et al. 2012). Further, using parabiotic mice, these authors demonstrated that after localized vaccinia virus skin infection, long-lived non-recirculating CD8+ T cells, termed T resident memory (T_{RM}) cells, are generated which provide superior protection over T_{CM} against re-infection (Jiang et al. 2012). These recent studies are in agreement with earlier observations that after parabiosis, CD8+ memory T cells in brain and intestinal mucosa did not equilibrate for weeks to months, indicating the presence of stable populations of memory T cells (Klonowski et al. 2004).

The factors that govern the residence vs. recirculation capacity are not completely delineated. Effector T cells downregulate CCR7, whereas central memory T cells retain it, and the available evidence suggests that CCR7 is the pivotal chemokine receptor for regulating the recruitment of cells toward the lymphatic vessel, including for T cells (Bromley et al. 2005; Debes et al. 2005). In chronic inflammation, however, other receptors besides CCR7 are able to mediate lymphocyte trafficking into lymph (Brown et al. 2010). In addition, the general motility of memory CD4+ T lymphocytes in skin is greater than that of memory CD8+ T lymphocytes (Gebhardt et al. 2011), possibly relating to why the former are more prone to egress through lymphatics.

To further expound upon the selectivity of cells that do and do not recirculate through afferent lymphatics, a recent study adopted the photo-switchable Kaede mouse (cells fluoresce green until exposure to UV light when they fluoresce red) to show that during a cutaneous immune response, more than half of the T cells moving from the skin to the draining lymph nodes are regulatory T cells (Tomura et al. 2010) (Fig. 4.1). This result seems to some extent to undermine the idea that cells recycling through lymph are there to survey tissues for antigens and then depart if all is well. While that concept is likely partly true, the prominent presence of T regulatory cells in lymph raises the possibility that recirculation of cells through tissues and their exit through lymph has a particular regulatory function that keeps immune responses appropriately dampened during homeostasis and perhaps could play an important role in the maintenance of tolerance to self antigen.

4.4 Dendritic Cells in Afferent Lymph

Dendritic cells (DCs), mononuclear phagocytes specialized for the regulation of adaptive immunity, are a distinct lineage of cells that develop using different precursors and transcriptional programs than macrophages (Gautier et al. 2012; Miller et al. 2012). DCs are dependent upon the cytokine flt3 for development and maintenance and are capable of trafficking through afferent lymph, whereas resident macrophages are rare to absent in lymph (Fig. 4.1). Indeed, highlighting their sessile nature, macrophages do not appear to arise from monocytes as long thought but maintain themselves locally in tissues after they are seeded from yolk sac macrophages (Schulz et al. 2012).

In the skin of mice, there are five DC subsets (Henri et al. 2010) that have the capacity to mobilize to lymph nodes through lymphatics. They are distinguished by distinct expression of an array of different C-type lectins and differential expression of markers like CD11b and CD103 (Fig. 4.1). Indeed, in all tissues, distinct expression of CD103 and CD11b comprise two major subsets of DCs that have differing biological properties. In the lung, CD103+ DCs are able to engulf apoptotic cells in contrast to CD11b+ DCs, and CD103+ DCs are associated with directing CD8+ T cell responses, whereas CD11b+ DCs appear to direct CD4+ cells (Eisenbarth et al. 2012). Remarkably, the induction of lymph-migrating potential in CD11b+ DCs in response to a range of stimuli required triggering of a common sensor, NLRP10, a Nod-like receptor (Eisenbarth et al. 2012). In the intestine, some studies suggested that CD103+ DCs were the sole subset in mouse afferent lymph draining the intestine (Schulz et al. 2009), but this finding has been contradicted (Cerovic et al. 2012), which is much more logical, as studies in rats for years have pointed to the heterogeneity of DCs in the afferent lymph, including cells that resembled and derived from monocytes (Yrlid et al. 2006). However, some DC subtypes, including plasmacytoid DCs that are quite abundant in lymph nodes (Diacovo et al. 2005) and are critical for type 1 IFN production and in tolerogenic responses, are not found in intestinal or hepatic lymph in the steady state or after TLR7/8 stimulation (Yrlid et al. 2006).

As mentioned earlier, much of what we know about the molecular events that control cell entry into lymphatic vessels has been learned from the study of DCs. Recent studies involving intravital imaging add new surprising findings that DCs adhere to the luminal surface of lymphatic vessels after entry into the vessels (Tal et al. 2011). This delay has the potential to alter the pace of DC emigration to lymph nodes and account for the arrival of different DC subsets in sequential fashion (Kissenpfennig et al. 2005), although it remains to be understood what such a regulatory event means for the immune response.

Why do DCs travel in lymph? In contrast to T lymphocytes, where the purpose of recirculation is unclear, the point of DC travel to the lymph node via lymphatic vessels has been thought to be well understood. The first concept was that they populate the lymph node through such a route, but it turns out that few lymph node DCs arrive through the afferent lymphatics (Jakubzick et al. 2008) and that most populate the lymph node through the arrival of DC precursors that traverse the HEV (Liu et al. 2009). A second concept was that those that do arrive through lymphatics, albeit typically a minority in the lymph node, are critical for carrying antigen to the lymph node for the purposes of inducing immunity. In general, data support this concept (Allan et al. 2006; Allenspach et al. 2008; Itano et al. 2003), but it is possible that data from CCR7$^{-/-}$ mice used to support this concept could be interpreted in other ways. Interestingly, DCs that arrive through afferent lymph participate in the organization of the lymph node itself and regulate the ability of HEV to recruit T cells (Wendland et al. 2011). Another reason to reflect on the possibility that immunologists still have much to learn about the importance of DC emigration to lymph nodes through lymphatics comes through the investigation of the overall gene expression profile of mouse lymph–migratory DCs, collected for gene expression analysis in the steady state (Miller et al. 2012). Migratory DCs recovered from the lymph node stood apart from the DC subsets collected either from the upstream tissue or as part of the lymph node resident DCs that enter by way of HEV, rather than maintaining similarity to their tissue counterparts. That is, CD103+CD11b– DCs that migrated from the lung to the lung-draining lymph node more closely clustered with the other major DC subset, the CD103–CD11b+ DCs that also emigrated from the lung, than they clustered with their own counterparts (CD103+ CD11b– DCs) that had not yet left the parenchyma of the lung. The change in gene expression was not solely a general shift in the lymph node environment because lymph migratory DCs did not cluster with lymph node DCs that arrived through the HEV. Part of the distinct set of genes that distinguishes lymph–migratory DCs are regulatory genes involved in suppressing T cell immune responses (Miller et al. 2012). It is unknown whether DCs that emigrate through lymphatics in the context of immunization or inflammation retain expression of these regulatory genes or if their expression is confined to resting lymph nodes. However, the findings fit nicely with a new study from the laboratory of Gordon Macpherson (Foster et al. 2012), whose career focused on direct cannulation and characterization of lymph DCs. MacPherson and coworkers have observed that migratory DCs retain key co-stimulatory molecules, such as CD40, in intracellular pools throughout migration in lymph, and only TCR engagement stimulates relocalization to the cell surface, whereas LPS was surprisingly ineffective. Thus, it appears that the migratory DC carries a regulatory phenotype, though it can be poised to activate T cells if the right signal is present. This observation would explain why most lymph–migratory DCs are recovered in a regulatory state in resting lymph nodes. Though somewhat surprising, this expression profile fits with other data that loss of migratory DCs leads not so much to an abrogation of adaptive immunity as to a loss of its regulatory control (Mori et al. 2001; Thomas et al. 2012) and even a reduced ability to induce tolerance (Thomas et al. 2012).

4.5 Trafficking into Efferent Lymph

The comparison of afferent and efferent lymph in sheep generated our understanding that naïve T cells are not recirculating through tissue but readily leave lymph nodes to enter efferent lymph. Thus, the concept of naïve T cells in afferent lymph seemed wrong and unsupported. However, Forster and colleagues have recently highlighted common misconceptions that have prevailed since these studies in sheep shaped our understanding of the cellular contents of lymph (Braun et al. 2011; Forster et al. 2012). The issue is that many lymph nodes are organized in chains, so the afferent lymphatic vessels of lymph nodes downstream of a given sentinel lymph node draining the parenchyma of an organ directly will receive efferent input from the upstream lymph node. Thus, significant naïve T cells travel in afferent lymph as they transit between lymph nodes in chains (Braun et al. 2011); only afferent lymphatic vessels that originate within and drain the parenchyma of organs, the type cannulated in most if not all studies of afferent lymph in sheep, are devoid of significant naïve T lymphocytes. Thus, though immunologists do not often consider it, the entry of recirculating naïve T cells into lymph nodes through afferent lymph is substantial. Interestingly, a recent study using a flank HSV inoculation model demonstrated that activated T cells egressing from the lymph node draining the primary site of infection could enter the next lymph node in the chain but would be confined to the sinusoidal network of that secondary lymph node, unless this lymph node was directly associated with the peripheral infection in which case the T cells could access the parenchyma of the lymph node (Eidsmo et al. 2012).

The control of lymphocyte egress from lymph nodes has received much attention and a critical mediator in this process is sphingosine-1-phosphate (S1P) (Cyster and Schwab 2012). S1P signals promote lymph node egress by actually competing with GPCR signals including CCR7, which aim to retain lymphocytes within lymph nodes (Pham et al. 2008). Thus T cells lacking CCR7 exit lymph nodes more rapidly than those expressing CCR7 (Pham et al. 2008). This contrasts with DCs, which are difficult to find in successive lymph nodes regardless of whether they express CCR7 (Braun et al. 2011). Indeed, the presence of DCs in efferent lymph is a far more contentious issue. It is widely thought that DCs that enter the lymph node from the periphery die locally (Kamath et al. 2002). However there are reports from the 1970/1980s that show 0.1 % of efferent lymph cells are macrophages, which we now know are likely to be primarily cells which we call DCs (Anderson et al. 1981; Bell 1979; De Martini et al. 1983). There is further evidence that DCs can enter efferent lymph, particularly after inflammatory stimuli. After epicutaneous application of both the chemical carcinogen 7,12-dimethylbenz(a)anthracene (DMBA) and the contact sensitizing antigen 2,4,6-trinitrochlorobenzene (TNCB), DC migration to lymph nodes is increased and DCs can be found in the efferent lymph following application of the former but not the latter (Dandie et al. 1994). Further, subcutaneous beryllium increases macrophage and DC numbers in the efferent lymph (Hall 1984). DCs can carry tracers first to the popliteal LN and then to a smaller extent to the iliac lymph node beyond (Angeli et al. 2006), and DCs injected in the skin can lead to priming of T cells in the spleen providing indirect evidence that DCs may reach the blood from tissues (Catalina et al. 1996). While there is some disagreement between studies, all studies point to a very low rate of transit for DCs into efferent lymph.

4.6 Concluding Remarks

Likely, we still have a lot to learn about the biological roles of lymph–migratory DCs and some of our presumptions about their roles may require revisiting. As discussed here, many recent works have challenged the two leading notions that prevailed only a decade ago: (1) that T cells recirculated to survey tissue for antigens and they leave through lymph if they don't find antigen; (2) the lymph node is populated with DCs because they travel into nodes from afferent lymph and nodes without lymph trafficking will not support induction of adaptive immune responses. Now, while these notions are not to be entirely erased, new data allow us to modify our understanding of the purpose of cell trafficking through afferent lymph. The retention of effector T cells in tissues without the need to recirculate through lymph indicates that immunity in peripheral organs and the survival of effector lymphocytes do not require that these cells revisit lymph nodes periodically. Instead, an unexpected theme has emerged that points to strong regulatory roles in lymph trafficking cells: the predominance of regulatory T cells in lymph, the regulatory state of DCs when they arrive in resting LNs, and the inability to maintain immunological tolerance in mouse strains that have abrogated cellular transit through lymph. It remains unknown what signals determine the disequilibrium in gaining access to lymph (Fig. 4.1). It is interesting that B cells unable to leave skin have the same phenotype as those unable to leave spleen. Likely, intrinsic features of certain cell phenotypes regulate their migratory potential through lymphatics, but at present these features are rather unclear. With the rapid pace of research on this topic, we can look forward to many new insights in the coming decade and beyond.

References

Allan RS, Waithman J, Bedoui S, Jones CM, Villadangos JA, Zhan Y, Lew AM, Shortman K, Heath WR, Carbone FR (2006) Migratory dendritic cells transfer antigen to a lymph node-resident dendritic cell population for efficient CTL priming. Immunity 25:153–162

Allenspach EJ, Lemos MP, Porrett PM, Turka LA, Laufer TM (2008) Migratory and lymphoid-resident dendritic cells cooperate to efficiently prime naive CD4 T cells. Immunity 29:795–806

Anderson AO, Warren JT, Gasser DL (1981) Presence of lymphoid dendritic cells in thoracic duct lymph from Lewis rats. Transplant Proc 13:1460–1468

Angeli V, Ginhoux F, Llodra J, Quemeneur L, Frenette PS, Skobe M, Jessberger R, Merad M, Randolph GJ (2006) B cell-driven lymphangiogenesis in inflamed lymph nodes enhances dendritic cell mobilization. Immunity 24:203–215

Baluk P, Fuxe J, Hashizume H, Romano T, Lashnits E, Butz S, Vestweber D, Corada M, Molendini C, Dejana E, McDonald DM (2007) Functionally specialized junctions between endothelial cells of lymphatic vessels. J Exp Med 204:2349–2362

Bell EB (1979) Antigen-laden cells in thoracic duct lymph. Implications for adoptive transfer experiments. Immunology 38:797–808

Braun A, Worbs T, Moschovakis GL, Halle S, Hoffmann K, Bolter J, Munk A, Forster R (2011) Afferent lymph-derived T cells and DCs use different chemokine receptor CCR7-dependent routes for entry into the lymph node and intranodal migration. Nat Immunol 12:879–887

Bromley SK, Thomas SY, Luster AD (2005) Chemokine receptor CCR7 guides T cell exit from peripheral tissues and entry into afferent lymphatics. Nat Immunol 6:895–901

Brown MN, Fintushel SR, Lee MH, Jennrich S, Geherin SA, Hay JB, Butcher EC, Debes GF (2010) Chemoattractant receptors and lymphocyte egress from extralymphoid tissue: changing requirements during the course of inflammation. J Immunol 185:4873–4882

Butcher EC, Picker LJ (1996) Lymphocyte homing and homeostasis. Science 272:60–66

Catalina MD, Carroll MC, Arizpe H, Takashima A, Estess P, Siegelman MH (1996) The route of antigen entry determines the requirement for L-selectin during immune responses. J Exp Med 184:2341–2351

Cerovic V, Houston SA, Scott CL, Aumeunier A, Yrlid U, Mowat AM, Milling SW (2012) Intestinal CD103(−) dendritic cells migrate in lymph and prime effector T cells. Mucosal Immunol 6(1):104–113

Clark RA, Watanabe R, Teague JE, Schlapbach C, Tawa MC, Adams N, Dorosario AA, Chaney KS, Cutler CS, Leboeuf NR, Carter JB, Fisher DC, Kupper TS (2012) Skin effector memory T cells do not recirculate and provide immune protection in alemtuzumab-treated CTCL patients. Sci Transl Med 4(117):117ra7

Cyster JG, Schwab SR (2012) Sphingosine-1-phosphate and lymphocyte egress from lymphoid organs. Annu Rev Immunol 30:69–94

Dandie GW, Watkins FY, Ragg SJ, Holloway PE, Muller HK (1994) The migration of Langerhans' cells into and out of lymph nodes draining normal, carcinogen and antigen-treated sheep skin. Immunol Cell Biol 72:79–86

De Martini JC, Fiscus SA, Pearson LD (1983) Macrophages in efferent lymph of sheep and their role in lectin-induced lymphocyte blastogenesis. Int Arch Allergy Appl Immunol 72:110–115

Debes GF, Arnold CN, Young AJ, Krautwald S, Lipp M, Hay JB, Butcher EC (2005) Chemokine receptor CCR7 required for T lymphocyte exit from peripheral tissues. Nat Immunol 6:889–894

Diacovo TG, Blasius AL, Mak TW, Cella M, Colonna M (2005) Adhesive mechanisms governing interferon-producing cell recruitment into lymph nodes. J Exp Med 202:687–696

Drexhage HA, Mullink H, de Groot J, Clarke J, Balfour BM (1979) A study of cells present in peripheral lymph of pigs with special reference to a type of cell resembling the langerhans cell. Cell Tissue Res 202:407–430

Eidsmo L, Stock AT, Heath WR, Bedoui S, Carbone FR (2012) Reactive murine lymph nodes uniquely permit parenchymal access for T cells that enter via the afferent lymphatics. J Pathol 226:806–813

Eisenbarth SC, Williams A, Colegio OR, Meng H, Strowig T, Rongvaux A, Henao-Mejia J, Thaiss CA, Joly S, Gonzalez DG, Xu L, Zenewicz LA, Haberman AM, Elinav E, Kleinstein SH, Sutterwala FS, Flavell RA (2012) NLRP10 is a NOD-like receptor essential to initiate adaptive immunity by dendritic cells. Nature 484:510–513

Forster R, Braun A, Worbs T (2012) Lymph node homing of T cells and dendritic cells via afferent lymphatics. Trends Immunol 33:271–280

Foster N, Turnbull EL, MacPherson GG (2012) Migrating lymph dendritic cells contain intracellular CD40 which is mobilized to the immunological synapse during interactions with antigen-specific T lymphocytes. J Immunol 189(12):5632–5637

Gautier EL, Shay T, Miller J, Greter M, Jakubzick C, Helft J, Chow A, Elpek KG, Gordonov S, Mazloom AR, Chua W-J, Hansen TH, Ma'ayan A, Turley SJ, Merad M, Randolph GJ (2012) Gene expression profiles and transcriptional regulatory pathways underlying murine tissue macrophage identity and diversity. Nat Immunol 13(11):1118–1128

Gebhardt T, Wakim LM, Eidsmo L, Reading PC, Heath WR, Carbone FR (2009) Memory T cells in nonlymphoid tissue that provide enhanced local immunity during infection with herpes simplex virus. Nat Immunol 10:524–530

Gebhardt T, Whitney PG, Zaid A, Mackay LK, Brooks AG, Heath WR, Carbone FR, Mueller SN (2011) Different patterns of peripheral migration by memory CD4+ and CD8+ T cells. Nature 477:216–219

Geherin SA, Fintushel SR, Lee MH, Wilson RP, Patel RT, Alt C, Young AJ, Hay JB, Debes GF (2012) The skin, a novel niche for recirculating B cells. J Immunol 188:6027–6035

Hall JG (1984) Studies on the adjuvant action of beryllium. I. Effects on individual lymph nodes. Immunology 53:105–113

Hein WR, McClure SJ, Miyasaka M (1987) Cellular composition of peripheral lymph and skin of sheep defined by monoclonal antibodies. Int Arch Allergy Appl Immunol 84:241–246

Henri S, Poulin LF, Tamoutounour S, Ardouin L, Guilliams M, de Bovis B, Devilard E, Viret C, Azukizawa H, Kissenpfennig A, Malissen B (2010) CD207+ CD103+ dermal dendritic cells cross-present keratinocyte-derived antigens irrespective of the presence of Langerhans cells. J Exp Med 207:189–206

Itano AA, McSorley SJ, Reinhardt RL, Ehst BD, Ingulli E, Rudensky AY, Jenkins MK (2003) Distinct dendritic cell populations sequentially present antigen to CD4 T cells and stimulate different aspects of cell-mediated immunity. Immunity 19:47–57

Jakubzick C, Bogunovic M, Bonito AJ, Kuan EL, Merad M, Randolph GJ (2008) Lymph-migrating, tissue-derived dendritic cells are minor constituents within steady-state lymph nodes. J Exp Med 205:2839–2850

Jennrich S, Lee MH, Lynn RC, Dewberry K, Debes GF (2012) Tissue exit: a novel control point in the accumulation of antigen-specific CD8 T cells in the influenza a virus-infected lung. J Virol 86:3436–3445

Jiang X, Clark RA, Liu L, Wagers AJ, Fuhlbrigge RC, Kupper TS (2012) Skin infection generates non-migratory memory CD8+ TRM cells providing global skin immunity. Nature 483:227–231

Kamath AT, Henri S, Battye F, Tough DF, Shortman K (2002) Developmental kinetics and lifespan of dendritic cells in mouse lymphoid organs. Blood 100:1734–1741

Kelly RH (1970) Localization of afferent lymph cells within the draining node during a primary immune response. Nature 227:510–513

Kissenpfennig A, Henri S, Dubois B, Laplace-Builhe C, Perrin P, Romani N, Tripp CH, Douillard P, Leserman L, Kaiserlian D, Saeland S, Davoust J, Malissen B (2005) Dynamics and function of Langerhans cells in vivo: dermal dendritic cells colonize lymph node areas distinct from slower migrating Langerhans cells. Immunity 22:643–654

Klonowski KD, Williams KJ, Marzo AL, Blair DA, Lingenheld EG, Lefrancois L (2004) Dynamics of blood-borne CD8 memory T cell migration in vivo. Immunity 20:551–562

Liu K, Victora GD, Schwickert TA, Guermonprez P, Meredith MM, Yao K, Chu FF, Randolph GJ, Rudensky AY, Nussenzweig M (2009) In vivo analysis of dendritic cell development and homeostasis. Science 324:392–397

Mackay CR, Kimpton WG, Brandon MR, Cahill RN (1988) Lymphocyte subsets show marked differences in their distribution between blood and the afferent and efferent lymph of peripheral lymph nodes. J Exp Med 167:1755–1765

Mackay CR, Marston WL, Dudler L (1990) Naive and memory T cells show distinct pathways of lymphocyte recirculation. J Exp Med 171:801–817

MartIn-Fontecha A, Sebastiani S, Hopken UE, Uguccioni M, Lipp M, Lanzavecchia A, Sallusto F (2003) Regulation of dendritic cell migration to the draining lymph node: impact on T lymphocyte traffic and priming. J Exp Med 198:615–621

Miller JC, Brown BD, Shay T, Gautier EL, Jojic V, Cohain A, Pandey G, Leboeuf M, Elpek KG, Helft J, Hashimoto D, Chow A, Price J, Greter M, Bogunovic M, Bellemare-Pelletier A, Frenette PS, Randolph GJ, Turley SJ, Merad M (2012) Deciphering the transcriptional network of the dendritic cell lineage. Nat Immunol 13:888–899

Mori S, Nakano H, Aritomi K, Wang CR, Gunn MD, Kakiuchi T (2001) Mice lacking expression of the chemokines CCL21-ser and CCL19 (plt mice) demonstrate delayed but enhanced T cell immune responses. J Exp Med 193:207–218

Olszewski WL, Grzelak I, Ziolkowska A, Engeset A (1995) Immune cell traffic from blood through the normal human skin to lymphatics. Clin Dermatol 13:473–483

Pflicke H, Sixt M (2009) Preformed portals facilitate dendritic cell entry into afferent lymphatic vessels. J Exp Med 206:2925–2935

Pham TH, Okada T, Matloubian M, Lo CG, Cyster JG (2008) S1P1 receptor signaling overrides retention mediated by G alpha i-coupled receptors to promote T cell egress. Immunity 28:122–133

Pugh CW, MacPherson GG, Steer HW (1983) Characterization of nonlymphoid cells derived from rat peripheral lymph. J Exp Med 157:1758–1779

Schulz O, Jaensson E, Persson EK, Liu X, Worbs T, Agace WW, Pabst O (2009) Intestinal CD103+, but not CX3CR1+, antigen sampling cells migrate in lymph and serve classical dendritic cell functions. J Exp Med 206:3101–3114

Schulz C, Gomez Perdiguero E, Chorro L, Szabo-Rogers H, Cagnard N, Kierdorf K, Prinz M, Wu B, Jacobsen SE, Pollard JW, Frampton J, Liu KJ, Geissmann F (2012) A lineage of myeloid cells independent of Myb and hematopoietic stem cells. Science 336:86–90

Smith JB, McIntosh GH, Morris B (1970) The traffic of cells through tissues: a study of peripheral lymph in sheep. J Anat 107:87–100

Spry CJ, Pflug AJ, Janossy G, Humphrey JH (1980) Large mononuclear (veiled) cells like 'Ia-like' membrane antigens in human afferent lympn. Clin Exp Immunol 39:750–755

Steinman RM, Cohn ZA (1973) Identification of a novel cell type in peripheral lymphoid organs of mice. I. Morphology, quantitation, tissue distribution. J Exp Med 137:1142–1162

Tal O, Lim HY, Gurevich I, Milo I, Shipony Z, Ng LG, Angeli V, Shakhar G (2011) DC mobilization from the skin requires docking to immobilized CCL21 on lymphatic endothelium and intralymphatic crawling. J Exp Med 208:2141–2153

Thomas SN, Rutkowski JM, Pasquier M, Kuan EL, Alitalo K, Randolph GJ, Swartz MA (2012) Impaired humoral immunity and tolerance in K14-VEGFR-3-Ig mice that lack dermal lymphatic drainage. J Immunol 189:2181–2190

Tomura M, Honda T, Tanizaki H, Otsuka A, Egawa G, Tokura Y, Waldmann H, Hori S, Cyster JG, Watanabe T, Miyachi Y, Kanagawa O, Kabashima K (2010) Activated regulatory T cells are the major T cell type emigrating from the skin during a cutaneous immune response in mice. J Clin Invest 120:883–893

Wendland M, Willenzon S, Kocks J, Davalos-Misslitz AC, Hammerschmidt SI, Schumann K, Kremmer E, Sixt M, Hoffmeyer A, Pabst O, Forster R (2011) Lymph node T cell homeostasis relies on steady state homing of dendritic cells. Immunity 35:945–957

Yrlid U, Cerovic V, Milling S, Jenkins CD, Zhang J, Crocker PR, Klavinskis LS, MacPherson GG (2006) Plasmacytoid dendritic cells do not migrate in intestinal or hepatic lymph. J Immunol 177:6115–6121

Chapter 5
The Lymph Proteome, Peptidome, and Degradome

Cristina C. Clement and Laura Santambrogio

Abstract During the last decade advanced proteomic analysis have allowed detailed characterization of the composition of the lymphatic fluid. A picture start to emerge indicating that far from being an ultra filtrate of the plasma the lymph carries a rich repertoire of proteins and peptides reflecting the tissue of origin and its physiological state. Indeed in few pathological conditions it has been shown that proteins transported by the lymph closely represent a molecular signature of the diseased organs. Altogether the "omic analysis" (proteomic, peptidomic, degradomic and lipidomic) prove to be a powerful tool to fingerprint variations in the lymph fluid collected from different parenchymal organs and to discover molecular biomarkers for the early detection of diseases.

5.1 Introduction

Physiological and pathological conditions in parenchymal organs are associated with protein expression profiles whose molecular signatures are often reflected in the fluids bathing tissue/organs. A major challenge of clinical chemistry is to use modern proteomic and peptidomic technologies to qualitatively and quantitatively map these protein profiles in order to provide molecular fingerprints of healthy or diseased conditions. Proteomic approaches can be indeed utilized to discover molecular biomarkers for the early detection of diseases, monitoring disease progression, and assessing the efficacy of therapeutic regimens (Meng and Veenstra 2007, 2011; Veenstra et al. 2005; Zimmerman et al. 2012; Qian et al. 2006). To date, many proteomic analyses of human body fluids, including plasma/serum, urine, cerebrospinal fluid, saliva, bronchoalveolar lavage, synovial fluid, nipple aspirate

C.C. Clement (✉) • L. Santambrogio
Department of Pathology, Albert Einstein College of Medicine, 1300 Morris Park Avenue, New York, NY, USA
e-mail: cristina.clement@einstein.yu.edu

L. Santambrogio (ed.), *Immunology of the Lymphatic System*,
DOI 10.1007/978-1-4614-3235-7_5, © Springer Science+Business Media New York 2013

fluid, tear fluid, lymph, and amniotic fluid have been performed (Meng and Veenstra 2007; Veenstra et al. 2005; Qian et al. 2006). However, to date the most comprehensive proteomic analysis has been performed on easily collectable biofluids, mostly plasma/serum and urine. In contrast, the composition of rodent and human lymph has been elusive, until very recently, due to the technical difficulty in collecting samples from the lymphatic capillaries (Veenstra et al. 2005; Qian et al. 2006).

In this chapter we will present a comprehensive overview of the proteomic and peptidomic composition of lymph as determined by advanced mass spectroscopy technologies performed by several groups. Since many of the lymph protein result from plasma ultrafiltration, the lymph protein expression profile will also be compared with that of plasma.

5.2 Lymph Formation

The pre-nodal afferent lymph is the fluid which directly derives from the extracellular milieu from every parenchymal organ and contains products derived from the organ metabolism/catabolism (Meng and Veenstra 2007; Interewicz et al. 2004; Clement et al. 2010, 2011). The detailed physiological studies on lymph circulation highlights the collection of this biological fluid into open end lymphatic capillaries, which form a mesh-like network distributed throughout the body (Clement et al. 2011). The lymph then enters through the cortical area of the node and conveys a representative sampling of the interstitial fluid to the nodal antigen-presenting cells before exiting through the efferent lymphatic vessels, or the central vein located in the nodal sinus (Clement et al. 2010, 2011).

Original work performed by Olsewski's and Miller's laboratories (Interewicz et al. 2004; Olszewski et al. 2001; Nanjee et al. 2000) emphasizes that the actual composition of the lymph is dictated by the blood capillary filtration rate, permeability of the capillary walls, metabolic state of parenchymal cells, and draining ability of the lymphatic capillaries. All these factors are constantly changing, depending on the physiological or pathological condition of the tissue or organ from which the lymph is drained (Interewicz et al. 2004; Olszewski et al. 2001). Indeed, in contrast to blood which is sampled from a large vein in the general circulation, the lymph is collected from the subcutaneous capillaries. Thus, the proteomic and peptidomic analysis of the lymph would reflect more closely the metabolism from which the capillaries drain (Olszewski 1991). For example, it was postulated that the composition of the lymph collected in a foot capillary, would be slightly different in resting conditions or during exercise, or cold and warm temperatures, the position of the leg extremity with respect to the heart level, and pathological skin processes. Similarly, the luminal content of the gut would influence intestinal lymph composition; hyperpnea would increase the lung lymph flow and change the chemical content of tissue fluids in the interstitial compartment and thus the peptidomic/proteomic composition of the lymph (Olszewski 1991). The new development and optimization of much more sensitive mass spectrometry-coupled assays will aid the discovery and validation of lymph biomarkers under physiological and pathological conditions.

5.3 Lymph and Plasma Proteome

Over the years very comprehensive databases have been compiled for human and rodent plasma proteins under physiological and pathological conditions (Tannu and Hemby 2006; Shen et al. 2006; Echan et al. 2005; Ahn and Simpson 2007; Armandola 2003; Whiteaker et al. 2007; Anderson et al. 2004; Zhou et al. 2004). So far, much less information is available for the lymph, mostly due to the difficulty in collecting the samples. Until a decade ago, the dominating notion was that the proteomic profile of pre-nodal afferent lymph would mostly overlap with that of plasma, since the former was considered to be an ultrafiltrate of the latter, although with a protein concentration anticipated to be lower in lymph compared to plasma. However, this notion was challenged by a few comparative analyses of lymph and plasma, which reported the presence of specific proteins in one fluid but not the other (Leak et al. 2004).

The first studies on lymph proteomics were performed on ovine lymph, since the size of the lymphatic capillaries allowed easy cannulation and enough material was retrieved for the analysis (Interewicz et al. 2004; Olszewski et al. 2001; Leak et al. 2004). Leak et al. (2004) described the first analysis of the pre-nodal, afferent ovine lymph and compared it with the matched plasma from the same ovine source. Their results, obtained by SELDI-TOF-MS and 2D-PAGE, demonstrated for the first time some relative differences in protein expression profiles between normal sheep lymph and plasma. Major proteins found in lymph, but also at lower concentrations in plasma, included fibrinogen alpha- and beta-chains, immunoglobulin (IgG), serotransferrin precursor, lactoferrin, and apolipoprotein A-1. Two proteins that were identified as uniquely expressed in lymph were glial fibrillary acidic protein and neutrophil cytosol factor-1 (Leak et al. 2004). The presence of the protease inhibitor, a-antiproteinase, demonstrated that the lymph also contains enzymatic components, confirming earlier studies which showed that the protease activator (tPA) and its inhibitor (PAI-1) were secreted by the lymphatic endothelium and were also present in the lymph (Leak et al. 2004).

In another independent study, performed by Goldfinch et al. (2008) ovine gastric lymph has been used to monitor the proteomic changes occurring in tissue fluid in response to infection with the parasitic nematode, *Teladorsagia circumcincta*. In this research, lymph was collected temporally over an experimental infection period and subjected to 2DE and proteomic analysis using MALDI-TOF MS and MS/MS. One important feature of the sample preparation was the immunodepletion of the lymph samples from albumin and IgG using immunoaffinity columns (Echan et al. 2005). Goldfinch et al. (2008) were able to identify a number of proteins by MS which had not previously been identified. These included multiple charged forms of alpha-1 acid glycoprotein and alpha-1 beta glycoprotein, two distinct forms of serum amyloid A and apolipoprotein A-1, and actin. The main serum/plasma proteins already identified by other groups, using 2-DE methodology (Whiteaker et al. 2007; Anderson et al. 2004), were also found in the gastric ovine lymph, including fibrinogen beta and gamma chains, alpha-2-HS-glycoprotein, haptoglobins, some classes of apolipoproteins (AI, AIV), transferrin, zinc-alpha-2 glycoprotein, C-type lectin, plasminogen, retinol-binding protein, and alpha-1 antitrypsin (Goldfinch et al. 2008). The studies of Goldfinch et al. outlined for the first time a strong correlation between an inflammatory condition and the protein expression profile of the lymph. More importantly, the analysis of the gastric lymph proteome highlighted the first potential biomarkers associated with one particular type of infection (*T. circumcincta*). These biomarkers, which included gelsolin, alpha-1 beta glycoprotein, and hemopexin, were further validated by Western blot assays and indicated changes in concentration according to the phases of infection (Goldfinch et al. 2008). Thus, Goldfinch et al. demonstrated for the first time, that a semiquantitative analysis of the changes in the lymph proteome during the course of a parasitic infection could represent a reliable method for monitoring the stages of the disease (Leak et al. 2004; Gretz et al. 1996; Goldfinch et al. 2008).

One step further in the analysis of the lymph proteome was the characterization of the mesenteric lymph during acute pancreatitis by Mittal et al. (2009). The main hypothesis that originated the analysis of the protein expression profile of the mesenteric lymph was that the protein fraction of this biological fluid could contain different, yet uncharacterized, toxic factors during different phases of this critical illnesses (Mittal et al. 2008, 2009). An analogous hypothesis drove the analysis of the proteomic composition of the plasma in many traumatic and pathophysiological conditions (Qian et al. 2006). Mesenteric lymph was collected from control rats and rats with taurocholate-induced acute pancreatitis and subjected to proteomic analysis using iTRAQ and liquid chromatography-tandem mass spectrometry. Two hundred and forty-five proteins including 35 hypothetical proteins were identified in mesenteric lymph. Eight of the 245 proteins had a significant increase in their relative abundance in acute pancreatitis-conditioned mesenteric lymph. These eight protein hits were then classified by their cellular location and molecular function using the Gene Ontology classification system. Seven of the proteins were pancreatic enzymes; four peptidases (carboxypeptidase B1, chymotrypsinogen B, carboxypeptidase A2, and cationic trypsinogen), one ester hydrolase (pancreatic lipase), one endoribonuclease (ribonuclease), and the pancreatic amylase 2 (Mittal et al. 2009), and one was a cytosolic protein.

From proteomics perspective, the characterization of acute pancreatitis-conditioned mesenteric lymph represents a successful approach involving the combined methodologies of immunodepletion (of albumin and IgG) and state-of-the-art iTRAQ coupled with nano-LC-tandem MS/MS, which helped to reveal an increased number of proteins (245) and to characterize in a quantitative manner fold changes in selective biomarkers (Mittal et al. 2009).

Additional work on proteomic profiling of mesenteric lymph was performed in rat models of hemorrhagic shock by the group of Kirk C Hansen and Ernest E. Moore, from the University of Colorado Health Science Center, Denver, CO (Masuno et al. 2006; Jordan et al. 2007; Peltz et al. 2009; Zurawel et al. 2010; Dzieciatkowska et al. 2011) and from the group of Sen-Yung Hsieh, from the Clinical Proteomic Center, Chang-Gung University, Taoyuan, Taiwan (Fang et al. 2010).

Recently, Moore's group has used the lymph samples from 11 patients to map the human mesenteric lymph after hemorrhagic shock (Dzieciatkowska et al. 2011). The pre and post-shock lymph was loaded at equal protein concentrations on 2D gels, followed by standard trypsin digestion and protein identification with tandem mass spectrometry (MS/MS). In one study the authors demonstrated that gelsolin, a plasma protein that functions to depolymerize actin filaments and mediate the inflammatory response by binding proinflammatory lipids such as lysophosphatidic acid, sphingosine-1-phosphate, and phosphoinositides, changed in concentration in the mesenteric lymph after hemorrhagic shock. Clinically, reduced gelsolin concentrations have been associated with increased mortality in critically ill, trauma, and burn patients (Masuno et al. 2006; Jordan et al. 2007). Additional extensive animal work using mesenteric lymph has helped establish a mechanistic link between gut ischemia/reperfusion (I/R) and distant organ injury. The advanced proteomics profiling of the mesenteric lymph posthemorragic shock (Peltz et al. 2009; Zurawel et al. 2010; Fang et al. 2010) characterized several proteins upregulated at least twofold in the post-shock as compared with the pre-shock lymph including proteins released from tissue injury, depletion of coagulation factors, and evidence of hemolysis (Peltz et al. 2009). The most important discovery was that the components of coagulation system were depleted, whereas products of hemolysis were increased. In addition, the proteomics analysis from the group of Ernest E. Moore showed that the haptoglobin was decreased, which was proposed to be related to an early post-shock hemolytic process. Unexpectedly, several protective protease inhibitors including alpha-2-macroglobulin, alpha-1-inhibitor 3, and inter-alpha-inhibitor H3 and H4, were found to be decreased. These findings suggested that the loss of protective protease inhibitors could be correlated with subsequent higher protease activity and the development of post-shock inflammation (Jordan et al. 2007; Peltz et al. 2009; Zurawel et al. 2010).

Other proteomic analysis of the mesenteric lymph followed from the studies of the Ernest E. Moore group, to further map changes in the proteomic profile of this biological fluid in posthemorrhagic shock (Dzieciatkowska et al. 2011). The Hansen's group collected mesenteric lymph from 11 patients with either lymphatic injuries, or during spine reconstruction surgery, or immediately before organ donation. The lymph was analyzed by label-free proteomic analysis resulting in the characterization of a total of 477 proteins identified from the 11 subject's lymph samples with greater than 99 % confidence. The proteomic methodology didn't involve 2D-DIGE analysis as in previous work, but a combination of immune-affinity depletion (of albumin and IgG) and 1DEF-SDS-PAGE coupled with nano-LC-tandem MS/MS. Several of the identified proteins were markers of hemolysis, oxidative stress, matrix degradation, and general tissue damage. Out of the top 200 proteins of the dataset, several were not found in the 889 Human Plasma Proteome Project (HPPP) dataset (Farrah et al. 2011; Omenn et al. 2005). This finding was consistent with work by Leak et al. that showed ovine lymph contains plasma proteins and additional metabolic products of connective tissue, cells and proteins released from lymphatic endothelium (Leak et al. 2004). It is important to mention that most of the identified proteins in rodent lymph were also identified in human mesenteric lymph (Jordan et al. 2007; Peltz et al. 2009; Zurawel et al. 2010; Dzieciatkowska et al. 2011; Fang et al. 2010).

While the methods employed in many of the above reported studies allowed for the identification of several hundred lymph proteins, the compiled datasets are far from being definitive. Indeed, the limited sample availability and inherent sensitivity of the analytical platform precludes, at least so far, the identification of many bioactive components such as cytokines or cellular mediators that are usually expressed at very low levels.

Our group recently reported the proteomic analysis of the afferent human lymph as compared to the plasma using a combination of immune-affinity depletion methods (for albumin and IgG) coupled with 1D SDS-PAGE and 2D-DIGE together with nano-LC-LTQ-Orbitrap tandem MS/MS. The 253 positively identified proteins ($p < 0.05$) indicated a common proteome, shared by the two biological fluids (144 out of 253 proteins), which was significantly enriched with complement activation and blood coagulation components, transporters, and protease inhibitors. In addition, the enriched proteome of human lymph (72 proteins) consisted of products derived from the extracellular matrix, apoptosis, and cellular catabolism. In contrast, the enriched proteome of human plasma (37 proteins) comprised soluble molecules of the coagulation system and cell–cell signaling factors. This was the first analysis of the human lymph that established, in a semiquantitative manner, that the physiological lymph proteome is comprised of a series of proteins that derive directly from the catabolism of parenchymal cells and the extracellular matrix. (Clement et al. 2013).

5.4 Plasma Peptidome and Degradome

Over the last decade, new studies have identified the presence of a low-molecular weight serum proteome composed of fragments derived from protein processing and cleavage. The presence of this "degradome" was thought to provide insight into tissue-specific metabolism and disease pathogenesis/progression (Hood et al. 2005; Geho et al. 2006; Antwi et al. 2009; Zheng et al. 2009; Zou et al. 2011). The analysis of this degradome/peptidome was initially limited by the protein depletion strategies employed during sample preparation for proteomic assays, which made it difficult to retrieve the low-molecular weight protein/peptides which are often bound to major lymph proteins (such as albumin, retinol-binding protein, or vitamin D-binding protein) and, until few years ago, by the low resolution of mass spectrometric analysis (Hood et al. 2005; Geho et al. 2006). The strategy to retrieve the low-molecular weight serum peptidome and cleaved proteome was eventually improved and involved peptide separation by ultrafiltration, enzymatic digestion, and liquid chromatography coupled to electrospray ion trap tandem mass spectrometry (LC-ESI-MS/MS). While this approach uncovered a remarkable richness of protein fragments observed in the low-molecular weight proteome, the incorporation of the tryptic digestion step eliminated specific information about the peptide sequences that occur naturally in the plasma or serum (Geho et al. 2006; Zou et al. 2011).

In addition, the intrinsic person-to-person variability, which is the case of most human biological samples, complicated the discovery of new biomarkers. As a consequence, initial efforts to discover a human plasma/serum degradome/peptidome were shifted to experimental mouse models of human diseases because they represented a highly controllable experimental system (Hood et al. 2005).

One of the most comprehensive approaches applied to mapping of the low-molecular weight cleaved proteome and peptidome of mouse serum is highlighted by the work of Brian L. Hood and colleagues (2005). By using differential labeling for trypsin-mediated incorporation of the stable isotope of oxygen, ^{18}O, they compared the low-molecular weight serum proteome of tumor-bearing mice with control mice. The digests were combined, fractionated by strong cation exchange chromatography, and analyzed by nanoflow-reversed phase liquid chromatography coupled online with tandem mass spectrometry. The analysis resulted in the identification of 6,003 proteins containing at least a single, fully tryptic peptide. The low-molecular weight degradome/peptidome identified in the tumor-bearing mice was partially annotated using the GO (gene ontology) classifications for cellular compartmentalization. The GO annotations indicated that 22 % of the peptide fragments were derived from the turnover of extracellular proteins. At the same time, the GO classification predicts that only 14 % of the entire healthy mouse proteome encodes for extracellular proteins. Thus, the comparative analysis between the plasma degradome/peptidome in tumor and control mice indicated that about 8 % of the newly identified extracellular peptidome derived from a differential processing of the extracellular proteome in the tumor vs. the healthy control mice. The analysis also revealed that sensitive, accurate, and robust mass spectrometry technologies, aiming to sequence the low-molecular weight degradome/peptidome, can be used as reliable tools for the characterization of differential processing pathways involved in the proteome's turnover under physiological or pathological conditions (Hood et al. 2005).

Over the past few years several more degradomes have been mapped including serum, plasma, lymph, synovial fluid, urine, and cerebrospinal fluid (Shen et al. 2010a, b, 2011, 2012; Koomen et al. 2005). The human degradome has been shown to contain more than 500 proteases responsible for protein degradation and control of protein quality and functions (Shen et al. 2011, 2012). To date, all studies agree that in pathological conditions, the mapped peptidome/degradome of different biological fluids is enriched from the one found in healthy, physiological conditions (Shen et al. 2010b, 2011, 2012). Indeed, among the sequenced peptides, fragments derived from the most abundant proteins found in human plasma and serum have been shown to be processed differently by proteases highly active in different diseases, such as cancer, rheumatoid arthritis (RA), and osteoporosis (Shen et al. 2010b, 2011, 2012). As such, the associated degradome identified in pathological states, including tumor progression, invasion, and metastasis, has been proposed as a potential source of biomarkers for clinical diagnosis purposes (Shen et al. 2010b, 2011, 2012).

Another inclusive analysis of the plasma peptidome is from the group of Richard Smith, from the Pacific Northwest National Laboratory (Shen et al. 2010a). Their combined chromatographic methods and high resolution Fourier transform (FT) tandem MS/MS technologies highlighted the power of the chosen fractionation methods, as compared to the LC/MS/MS methodology, in mapping a high number of substrates of the degradome. The fractionation of human blood proteins using a combination of affinity and strong exchange chromatographic methods (AC)/(SEC) enabled depletion of the abundant proteins and enrichment of the small- to medium-sized blood plasma degradome/peptidome components. Additionally, the chromatographic fractionation strategies were used in conjunction with de novo sequencing for the identification of posttranslational modifications and mutations using high resolution liquid chromatography (HRLC)-FT-MS/MS. More than 200 peptides generated from 29 protein substrates were confidently identified from the plasma of a single healthy person (Shen et al. 2010a). The identified peptides had MWs in the range of 0.5–14 kDa and identifications were achieved with extremely low (near zero) false discovery rates through searching the IPI human protein database (Shen et al. 2010a). Alpha-1,2-mannosidase IA, alpha-1B-glycoprotein, alpha-2-antiplasmin, alpha-2-HS-glycoprotein, angiotensinogen, antithrombin III variant, apolipoprotein A-IV, beta-2 microglobulin, C1-esterase inhibitor, ceruloplasmin, complement C3, complement C4, complement C9, complement factor D, fibrinogen, gelsolin, inter-alpha-trypsin inhibitor H4, kininogen-(HMW), kininogen-(LMW), leucine-rich alpha-2-glycoprotein, multimerin-1, pigment epithelium-derived factor, plasma serine protease inhibitor, thymosin beta-4, and transthyretin were among the most abundant proteomic substrates. Some of the sequenced peptides were shown to be extensively modified posttranslationally, including acetylation, acetylhexosamine, amidation, cysteinylation, didehydro, oxidation, and pyro-glu, as the major posttranslational modifications (Shen et al. 2010a).

Combining different MS dissociation methods, including collision-induced dissociation (CID), high energy collision dissociation (HCD), and electron transfer dissociation (ETD) together with Fourier transform MS/MS, the group of Shen et al. identified >1,000 unique peptides with a low-peptide-level false discovery rate (FDRs) (e.g., 0.6 %).

The peptides were derived from the degradation of protein substrates of pooled plasma samples collected from early stage breast cancer patients (Shen et al. 2012). This analysis provided an improved basis for investigating disease-related peptidome components and further highlighted the advantages of the global analysis of the peptidome to identify specific protein substrate targets, characterize cleavage specificities of processed proteins substrates, and for diagnostic purposes (Shen et al. 2010a, b, 2011, 2012).

A major concern with the peptidome/degradome analysis is the limit of tryptic digests in revealing the cleavage sites of naturally processed peptides. This problem was overcome by using different fractionation procedures, which enabled the purification of naturally processed peptides with MW <5,000 Da. One characterization of naturally cleaved serum and plasma peptidome is highlighted by the LC

MALDI-Tof/Tof (MS/MS) studies of low-molecular weight proteome of heparinized plasma performed by John M. Koomen et al. in the research group of Ryuji Kobayashi (Koomen et al. 2005). In this analysis a combination of 96-well plate fractionation of plasma and LCMALDI-MS/MS were used for profiling and directly sequencing abundant peptides in the low-molecular weight plasma proteome. Approximately 250 unique peptides were detected, corresponding to approximately 20 proteins. The majority of the peptides were identified as fragments from fibrinogen, complement components, carrier proteins, and antiproteases. The cleavage patterns were consistent with those of known plasma proteases, including thrombin, plasmin, and complement proteins, followed by aminopeptidase and carboxypeptidase activity (Koomen et al. 2005). Altogether, the remarkable work of Koomen et al. outlines some mechanistic insights into the processing pathways generating the plasma degradome by molecular analysis of the cleavage sites of the peptidome (Koomen et al. 2005).

Another significant contribution to the discovery of new peptide markers in the low-molecular weight fraction of the human proteome was generated by Yusuke Kawashima et al. (2010). The research team from the Center for Disease Proteomics at the Kitasato University School of Science, Japan, developed a differential solubilization (DS) method to extract low-molecular weight proteins/peptides in serum. A combination of peptide extraction with denaturating agents (urea, thiourea) followed by ice-cold acetone precipitation, differential centrifugation, and trifluoroacetic acid (TFA 0.1 %) extraction of the final proteome/peptides allowed the development of a new extraction procedure characterized by good reproducibility and yield as compared to typical peptide-extraction methods such as organic solvent precipitation and ultrafiltration. Using the DS method combined with reverse-phase HPLC fractionation followed by MALDI-TOF-MS, they characterized more than 1,500 peptides from 1 μL of serum samples, including low-abundance peptides in the subnanomolar range and many peptides bound to carrier proteins such as albumin.

The results of Yusuke Kawashima et al. provided the evidence that serum peptide analyses based on the DS method could greatly contribute to the discovery of novel low-abundance biomarkers (Kawashima et al. 2010). The major protein hits identified were the family of apolipoproteins (AII, AIV, CI, CII, CIII, CIV, E precursors, haptoglobins, serum amyloid A-4 precursor, hemoglobin subunit alpha, beta, delta, clusterin precursor, transthyretin, alpha-1-antitrypsin precursor, leucine-rich alpha-2-glycoprotein precursor, pigment epithelium-derived factor precursor, antithrombin-III precursor, N-acetylmuramoyl-L-alanine amidase precursor, alpha-1B-glycoprotein precursor, serum albumin precursor, serotransferrin precursor, serum albumin precursor, Ig kappa chain C region, and Ig lambda chain C regions (Kawashima et al. 2010).

An independent, notable contribution for the characterization of the plasma peptidome was reported by Rustam Ziganshin et al. (2011), who developed a new method for desorption of low-molecular weight peptides from abundant blood proteins. The diluted blood serum was heated to 98 °C for 15 min in order to dissociate the low-molecular weight peptides from the most abundant blood proteins.

Blood plasma/serum fractionation was performed on magnetic beads with a functionalized surface followed by heating of the resultant fractions. This methodology significantly increased the number of peptides detected by MALDI–TOF MS and enhanced the general reproducibility of mass spectrometry profiles determined by LC–MS/MS using an Agilent 6520 Accurate-Mass Q-TOF (Ziganshin et al. 2011). The processed peptides covered the major classes of protein antigens from the plasma such as apolipoproteins (A-I, A-II, A-IV, B-100, C-I, C-II, C-III, E, clusterin (apo J)), coagulation factors (XII, XIIIa), complement system (C3, C4, factor B), fibrinogens (alpha chain), fibronectin, gelsolin, hemopexin, histidine-rich glycoprotein, inter-alpha-trypsin inhibitor heavy chain H4, kininogen-1, pigment epithelium-derived factor, plasma protease C1 inhibitor, plasma retinol-binding protein, plasminogen, prothrombin, serotransferrin, transthyretin, vitronectin, cytoplasmic actin, antithrombin-III, and alpha-HS glycoprotein (Ziganshin et al. 2011).

As a whole, the reported analysis on the low-molecular weight plasma peptidome/degradome indicated, as expected, that the major classes of plasma proteins could always be detected by the mass spectrometric analysis, independently of the method used for the peptidome extraction. Nevertheless, an additional peptidome/degradome derived from the low abundant intracellular and extracellular proteomes could be identified accurately only by using multiple chromatographic fractionation methods coupled with high resolution FT-MS/MS technologies (Shen et al. 2010a, b, 2011, 2012; Koomen et al. 2005; Kawashima et al. 2010; Ziganshin et al. 2011).

5.5 Lymph Peptidome and Degradome

Throughout the last decade (2002–2012), many research studies have been performed on human, bovine, rat, and goat lymph collected at different sites, thus providing the first glimpse into the proteomic composition of the lymphatic fluid. From early on a major question asked from the data analysis was how qualitatively and quantitatively different is the plasma/serum peptidome from the lymph one? (Shen et al. 2010a, b, 2011, 2012) Also, would the peptidome/degradome overlap with the proteins found in the lymph or would it reflect the tissue where the lymph originated? Moreover, would the cleavage site indicate a similar set of enzymes involved in the processing for the lymph and plasma peptidome or indicate a different set of proteases? Even though many of the questions are still unresolved, in the next sections of this chapter we will highlight the major advances in the qualitative and quantitative characterization of the lymph peptidome.

The first mapping of the peptidome transported by the human lymph was recently described by our laboratory at Albert Einstein College of Medicine (Clement et al. 2010, 2011). The primary hypothesis that initiated the research on the human lymph peptidome was that the lymph could potentially carry a wider antigenic repertoire than the plasma and could in turn be a richer source of tissue-specific antigens. Additionally, differently from plasma, the lymph could carry an enriched processed proteome/peptidome since it directly collects from the extracellular milieu where products of tissue catabolism, tissue remodeling, cellular apoptosis, and extracellular matrix processing are collected before being transported to the draining lymph nodes (Clement et al. 2010, 2011).

Due to the difficulty in obtaining sufficient material, the peptidomic analysis was originally performed on pooled sample of human lymph from 18 healthy donors. Peptides were purified by ultrafiltration through 5,000 Da molecular weight filters or by reverse phase HPLC or gel filtration and then sequenced using both nanospray LC-MS/MS on a LTQ linear ion trap mass spectrometer (LTQ) and an Orbitrap/Velos (Clement et al. 2010 and unpublished data). More than 300 sequences were identified as self-peptides derived from both intracellular and extracellular proteins, mapping the variety of catabolic products transported by the human lymph (Clement et al. 2010). As expected, a large fraction of the peptidome was comprised of processed proteins derived from extracellular matrix proteins, cell adhesion molecules, and plasma membrane/receptors or from enzymes involved in the processing of the substrates which gave rise to the reported peptidome (metalloproteases/peptidases) (Clement et al. 2010).

Interestingly, many of the partially processed intracellular proteins were derived from the cytosolic fraction, nucleus (transcription factors and regulator of gene expression), mitochondria, golgi, and endoplasmic reticulum, most probably from parenchymal cells undergoing apoptosis. Indeed apoptotic cells have been found in the human lymph (Olszewski 1991). Pathway analysis of the mapped peptidome/degradome confirmed that the most abundant networks related to the biological turnover of the extracellular matrix, followed by cellular metabolism.

When the peptidome profile was compared to previously published MHC class II eluted peptidome, it appeared that the lymph peptide repertoire was partially skewed toward peptides derived from extracellular matrix proteins, cadherins, and cleavable or soluble receptors (43 %) (Clement et al. 2010). Likely, this was related to the source of the lymph, which was collected from the foot, and reflected the protein composition of the tissues (skin, adipose tissue, connective, and muscle) (Clement et al. 2010). In the future it would be interesting to collect lymph draining different organs and compare the degradome/peptidome for tissue-specific antigens.

A quantitative analysis of the peptides established that at least some of them were present in the circulating lymph in nanomolar concentration, proving that an immunologically relevant amount of partially processed proteins and peptides are carried by the human lymph (Clement et al. 2010, 2011). A preliminary analysis using the DAVID-WS program, which includes the characterization of the possible antigen processing machinery responsible for peptide cleavage, indicated that 8 % from the lymph peptidome is processed by metalloproteases and other peptidases.

This analysis further suggests that, unlike the proteins processed by local or nodal APC (antigen-presenting cells), which mostly produce epitopes constrained by endosomal activity, the identified self-antigens in the lymph could be derived from a wider variety of processing pathways, including caspases involved in cellular apoptosis, ADAMs and other metalloproteinases involved in surface receptor editing, cytokine processing, and/or matrix remodeling (Clement et al. 2010).

In the future it would be important to map the degradome machinery involved in the generation of the lymph peptidome. This type of analysis would involve mapping of the protease cleavage sites for the previously reported peptidome (Clement et al. 2010), using a similar approach to the one reported for the plasma peptidome by Yunfeg Shen et al. (2010a, b, 2011, 2012) or by Kobayashi's group (Koomen et al. 2005).

5.6 Conclusions

Over the course of the past 10 years, enormous progress has been made in the analysis of lymph fluid. In particular, advances in proteomic techniques have enabled a more comprehensive characterization of the lymph proteomic, peptidomic, and degradomic content than has heretofore been possible. However, much remains to be done. For example, it will be important to collect capillary lymph from different tissues in order to have an inclusive analysis of the tissue-specific self antigens that are transported to the draining lymph nodes. Additionally, many low abundant proteins and tissue-specific self antigens still remain to be mapped. Furthermore, with respect to the characterization of the lymph degradome, an analysis of the processing pathways that generate the previously reported peptidomes will also be an important endeavor, since it will specify the role of the over 500 mapped proteases in the human lymph and plasma that are involved in processing. Finally, since the lymph fluid carries the proteomic signature of the tissue that it drains, the immunological role of this tissue-specific proteome in maintenance of peripheral tolerance will also be an important question to answer.

References

Ahn S-M, Simpson RJ (2007) Body fluid proteomics: prospects for biomarker discovery. Proteomics Clin Appl 1(9):1004–1015

Anderson NL, Polanski M, Pieper R, Gatlin T, Tirumalai RS, Conrads TP, Veenstra TD, Adkins JN, Pounds JG, Fagan R, Lobley A (2004) The human plasma proteome: a nonredundant list developed by combination of four separate sources. Mol Cell Proteomics 3:311–326

Antwi K, Hostetter G, Demeure MJ, Katchman BA, Decker GA, Ruiz Y, Sielaff TD, Koep LJ, Lake DF (2009) Analysis of the plasma peptidome from pancreas cancer patients connects a peptide in plasma to overexpression of the parent protein in tumors. J Proteome Res 8(10):4722–4731

Armandola EA (2003) Proteome profiling in body fluids and in cancer cell signaling. MedGenMed 5(2):18

Clement CC, Cannizzo ES, Nastke MD, Sahu R, Olszewski W, Miller NE, Stern LJ, Santambrogio L (2010) An expanded self-antigen peptidome is carried by the human lymph as compared to the plasma. PLoS One 5(3):e9863

Clement CC, Rotzschke O, Santambrogio L (2011) The lymph as a pool of self-antigens. Trends Immunol 32(1):6–11

Clement CC, Aphkhazava D, Nieves E, Callaway M, Olszewsky W, Rotzschke O, Santambrogio L (2013) Protein expression profiles of human lymph and plasma mapped by 2D-Dige and 1D SDSpage coupled with nanolc-ESI-MS/MS bottom up proteomics. J Proteomics 78:172–187

Dzieciatkowska M, Wohlauer MV, Moore EE, Damle S, Peltz E, Campsen J, Kelher M, Silliman C, Banerjee A, Hansen KC (2011) Proteomic analysis of human mesenteric lymph. Shock 35(4):331–338

Echan LA, Tang HY, Li-Khan N, Lee K, Speicher DW (2005) Depletion of multiple high-abundance proteins improves protein profiling capacities of human serum and plasma. Proteomics 13:3292–3303

Fang JF, Shih LY, Yuan KC, Fang KY, Hwang TL, Hsieh SY (2010) Proteomic analysis of post-hemorrhagic shock mesenteric lymph. Shock 34(3):291–298

Farrah T, Deutsch EW, Omenn GS, Campbell DS, Sun Z, Bletz JA, Mallick P, Katz JE, Malmstrom J, Ossola R, Watts JD, Lin B, Zhang H, Moritz RL, Aebersold RH (2011) A high confidence human plasma proteome reference set with estimated concentrations in PeptideAtlas. Mol Cell Proteomics 10(9):M110.006353

Geho DH, Liotta LA, Petricoin EF, Zhao W, Araujo RP (2006) The amplified peptidome: the new treasure chest of candidate biomarkers. Curr Opin Chem Biol 10(1):50–55

Goldfinch GM, Smith WD, Imrie L, McLean K, Inglis NF, Pemberton AD (2008) The proteome of gastric lymph in normal and nematode infected sheep. Proteomics 8(9):1909–1918

Gretz JE, Kaldjian EP, Anderson AO, Shaw S (1996) Sophisticated strategies for information encounter in the lymph node: the reticular network as a conduit of soluble information and a highway for cell traffic. J Immunol 157:495–499

Hood BL, Lucas DA, Kim G, Chan KC, Blonder J, Issaq HJ, Veenstra TD, Conrads TP, Pollet I, Karsan A (2005) Quantitative analysis of the low molecular weight serum proteome using 18O stable isotope labeling in a lung tumor xenograft mouse model. J Am Soc Mass Spectrom 16(8):1221–1230

Interewicz B, Olszewski WL, Leak LV, Petricoin EF, Liotta LA (2004) Profiling of normal human leg lymph proteins using the 2-D electrophoresis and SELDI-TOF mass spectrophotometry approach. Lymphology 37:65–72

Jordan JR, Moore EE, Damle SS, Eckels P, Johnson JL, Roach JP, Redzic JS, Hansen KC, Banerjee A (2007) Gelsolin is depleted in post-shock mesenteric lymph. J Surg Res 143(1):130–135

Kawashima Y, Fukutomi T, Tomonaga T, Takahashi H, Nomura F, Maeda T, Kodera Y (2010) High-yield peptide-extraction method for the discovery of subnanomolar biomarkers from small serum samples. J Proteome Res 9(4):1694–1705

Koomen JM, Li D, Xiao LC, Liu TC, Coombes KR, Abbruzzese J, Kobayashi R (2005) Direct tandem mass spectrometry reveals limitations in protein profiling experiments for plasma biomarker discovery. J Proteome Res 4(3):972–981

Leak LV, Liotta LA, Krutzsch H, Jones M, Fusaro VS, Ross SJ, Zhao Y, Petricoin EF III (2004) Proteomic analysis of lymph. Proteomics 4:753–765

Masuno T, Moore EE, Cheng AM, Sarin EL, Banerjee A (2006) Bioactivity of postshock mesenteric lymph depends on the depth and duration of hemorrhagic shock. Shock 26(3):285–289

Meng Z, Veenstra TD (2007) Proteomic analysis of serum, plasma, and lymph for the identification of biomarkers. Proteomics Clin Appl 1(8):747–757

Meng Z, Veenstra TD (2011) Targeted mass spectrometry approaches for protein biomarker verification. J Proteomics 74(12):2650–2659

Mittal A, Middleditch M, Ruggiero K, Buchanan CM, Jullig M, Loveday B, Cooper GJ, Windsor JA, Phillips AR (2008) The proteome of rodent mesenteric lymph. Am J Physiol Gastrointest Liver Physiol 295(5):G895–G903

Mittal A, Phillips AR, Middleditch M, Ruggiero K, Loveday B, Delahunt B, Cooper GJ, Windsor JA (2009) The proteome of mesenteric lymph during acute pancreatitis and implications for treatment. JOP 10(2):130–142

Nanjee MN, Cooke CJ, Olszewski WL, Miller NL (2000) Lipid and apolipoprotein concentrations in prenodal leg lymph of fasted humans. Associations with plasma concentrations in normal subjects, lipoprotein lipase deficiency, and LCAT deficiency. J Lipid Res 41:1317–1327

Olszewski WL (1991) Lymph stasis: pathophysiology, diagnosis and treatment. CRC, Boca Raton, FL, pp 235–280 (Chaps. 10–11)

Olszewski WL, Pazdur J, Kubasiewicz E, Zaleska M, Cooke CJ, Miller NE (2001) Lymph draining from foot joints in rheumatoid arthritis provides insight into local cytokine and chemokine production and transport to lymph nodes. Arthritis Rheum 44:541–549

Omenn GS, States DJ, Adamski M, Blackwell TW, Menon R, Hermjakob H, Apweiler R, Haab BB, Simpson RJ, Eddes JS, Kapp EA, Moritz RL, Chan DW, Rai AJ, Admon A, Aebersold R, Eng J, Hancock WS, Hefta SA, Meyer H, Paik YK, Yoo JS, Ping P, Pounds J, Adkins J, Qian X, Wang R, Wasinger V, Wu CY, Zhao X, Zeng R, Archakov A, Tsugita A, Beer I, Pandey A, Pisano M, Andrews P, Tammen H, Speicher DW, Hanash SM (2005) Overview of the HUPO Plasma Proteome Project: Results from the pilot phase with 35 collaborating laboratories and multiple analytical groups, generating a core dataset of 3020 proteins and a publicly-available database. Proteomics 13:3226–3245

Peltz ED, Moore EE, Zurawel AA, Jordan JR, Damle SS, Redzic JS, Masuno T, Eun J, Hansen KC, Banerjee A (2009) Proteome and system ontology of hemorrhagic shock: exploring early constitutive changes in postshock mesenteric lymph. Surgery 146(2):347–357

Qian WJ, Jacobs JM, Liu T, Camp DG II, Smith RD (2006) Advances and challenges in liquid chromatography-mass spectrometry-based proteomics profiling for clinical applications. Mol Cell Proteomics 5(10):1727–1744

Shen Z, Want EJ, Chen W, Keating W, Nussbaumer W, Moore R, Gentle TM, Siuzdak G (2006) Sepsis plasma protein profiling with immunodepletion, three-dimensional liquid chromatography tandem mass spectrometry, and spectrum counting. J Proteome Res 5:3154–3160

Shen Y, Liu T, Tolić N, Petritis BO, Zhao R, Moore RJ, Purvine SO, Camp DG, Smith RD (2010a) Strategy for degradomic-peptidomic analysis of human blood plasma. J Proteome Res 9(5):2339–2346

Shen Y, Tolić N, Liu T, Zhao R, Petritis BO, Gritsenko MA, Camp DG, Moore RJ, Purvine SO, Esteva FJ, Smith RD (2010b) Blood peptidome-degradome profile of breast cancer. PLoS One 5(10):e13133

Shen Y, Toli N, Xie F, Zhao R, Purvine SO, Schepmoes AA, Moore RJ, Anderson GA, Smith RD (2011) Effectiveness of CID, HCD, and ETD with FT MS/MS for degradomic-peptidomic analysis: comparison of peptide identification methods. J Proteome Res 10(9):3929–3943

Shen Y, Toli N, Purvine SO, Smith RD (2012) Improving collision induced dissociation (CID), high energy collision dissociation (HCD), and electron transfer dissociation (ETD) Fourier transform MS/MS degradome-peptidome identifications using high accuracy mass information. J Proteome Res 11(2):668–677

Tannu NS, Hemby SE (2006) Two-dimensional fluorescence difference gel electrophoresis for comparative proteomics profiling. Nat Protoc 1:1732–1742

Veenstra TD, Conrads TP, Hood BL, Avellino AM, Ellenbogen RG, Morrison RS (2005) Biomarkers: mining the biofluid proteome. Mol Cell Proteomics 4(4):409–418

Whiteaker JR, Zhang H, Eng JK, Fang R, Piening BD, Feng LC, Lorentzen TD, Schoenherr RM, Keane JF, Holzman T, Fitzgibbon M, Lin C, Zhang H, Cooke K, Liu T, Camp DG II, Anderson L, Watts J, Smith RD, McIntosh MW, Paulovich AG (2007) Head-to-head comparison of serum fractionation techniques. J Proteome Res 6:828–836

Zheng X, Wu SL, Hincapie M, Hancock WS (2009) Study of the human plasma proteome of rheumatoid arthritis. J Chromatogr A 1216(16):3538–3545

Zhou M, Lucas DA, Chan KC, Issaq HJ, Petricoin E III, Liotta LA, Veenstra TD, Conrads TP (2004) An investigation into the human serum "interactome". Electrophoresis 25(9):1289–1298

Ziganshin R, Arapidi G, Azarkin I, Zaryadieva E, Alexeev D, Govorun V, Ivanov V (2011) New method for peptide desorption from abundant blood proteins for plasma/serum peptidome analyses by mass spectrometry. J Proteomics 74(5):595–606

Zimmerman LJ, Li M, Yarbrough WG, Slebos RJ, Liebler DC (2012) Global stability of plasma proteomes for mass spectrometry-based analyses. Mol Cell Proteomics 11(6):M111.014340

Zou X, Zhong L, Liu D, Yang B, Lou Y, Peng J, Rainer M, Feuerstein I, Muhammad NU, Huck CW, Bonn GK, Yin Y (2011) Novel multifunctional chitosan-GMA-IDA-Cu(II) nanospheres for high dynamic range characterization of the human plasma proteome. Anal Bioanal Chem 400(3):747–756

Zurawel A, Moore EE, Peltz ED, Jordan JR, Da mle S, Dzieciatkowska M, Banerjee A, Hansen KC (2010) Proteomic profiling of the mesenteric lymph after hemorrhagic shock: differential gel electrophoresis and mass spectrometry analysis. Clin Proteomics 8(1):1

Chapter 6
Aging and Lymphatic Contractility: Current Status

Anatoliy A. Gashev and Victor Chatterjee

Abstract Aging impairs lymph flow, which is crucial for fluid and macromolecule homeostasis, fat absorption, and immunity. Aging of the lymphatic system is grossly understudied and may be contributory to many of the aging-related diseases of the elderly. We evaluated aging-associated changes in adult (9 months) and aged (24 months) isolated rat thoracic ducts (TDs) and mesenteric lymphatic vessels (MLV) and demonstrated that there was decreased amplitude, frequency, tone and pumping in aged TDs and MLVs as compared to their adult counterparts. Differences in contractile behavior of aged isolated MLV and those observed in situ suggest an important regulatory involvement of some cellular elements in the immediate environment of aged MLV. We identified mast cells lining MLV and demonstrated a greater degree of basal activation of mast cells in aged mesentery. We also identified increased superoxide and peroxynitrite radicals and other signs of oxidative stress in aged MLV as compared to adult. These studies provided first ground for understanding of some of the mechanisms responsible for impairment of lymphatic function associated with aging.

The major task of the lymphatic system is lymph transport. This system of initial capillaries, transporting vessels, and nodes is designed to transport fluid, soluble molecules, and immune cells from the interstitium through the lymph nodes to the central veins. This system also provides the transportation route for inflammatory mediators, the products of tissue injury/destruction, foreign substances,

A.A. Gashev (✉)
Department of Medical physiology, College of Medicine Texas A&M Health Science Center, 702 SW H.K. Dodgen Loop, Temple, TX 76504, USA
e-mail: gashev@tamu.edu

V. Chatterjee
Department of Systems Biology and Translational Medicine, College of Medicine Texas A&M Health Science Center, 702 SW H.K. Dodgen Loop, Temple, TX 76504, USA

L. Santambrogio (ed.), *Immunology of the Lymphatic System*,
DOI 10.1007/978-1-4614-3235-7_6,

and tumor cells. Investigations over the last few decades have provided knowledge of lymph transport from lymphatic capillaries to the thoracic duct (TD) (Mislin and Schipp 1966; Hargens and Zweifach 1977; Unthank and Bohlen 1988; Schmid-Schonbein 1990; McHale 1990, 1992; Benoit and Zawieja 1994; Gashev and Zawieja 2001; von der Weid 2001; Gashev 2002; Bridenbaugh et al. 2003). There are several motive forces, which drive lymph centripetally. Traditionally these forces are divided into two groups depending on their origin (Gasheva et al. 2006). The term "active" or "intrinsic" lymph pump describes the lymph driving force that is generated by the active spontaneous contractions of the lymphangions (functional unit of contractile lymphatics consisting of the lymphatic valve and lymphatic wall up to the next downstream valve) (Mislin 1961). The contraction energy of lymphangions is essential to transport lymph against the opposing net hydrostatic gradient. Since the driving force of one such small pump is typically not enough to propel lymph all the way down to the central end of the lymphatic system, lymphatic vessels are organized in chains of these pumps. The term "passive" or "extrinsic" lymph pump combines together all other forces outside of the lymphatic vessel that are not dependent on the active contractions of lymphatic wall but may assist lymph flow in different regions of body. These "passive" forces include the influences of lymph formation, contractions of adjacent skeletal muscles, vasomotion of adjacent blood vessels, gastrointestinal motility, and breathing. Thus lymph flow occurs as a result of complicated combinations of the predominant active and occasional/regionally supportive passive lymph driving forces.

There is no evidence in any regional lymphatic network that phasic contractions of lymphatics occur simultaneously along the length of an entire lymphatic chain. Contrary to this idea, numerous reports have demonstrated the propagation of contractile waves along lymphatic vessels (McHale and Roddie 1976; McHale and Meharg 1992; Zawieja et al. 1993; Crowe et al. 1997). In some regional lymphatic networks (in lower limbs), an interrupted fluid column was demonstrated during the normal contractile activity of lymphangions (Armenio et al. 1981a, b; Gashev et al. 1990). In that situation, adjacent lymphangions work in a counter-phase fashion, with some lymphangions contracting while the next one or several lymphangions are in diastolic relaxation (Gashev et al. 1990). Thus each lymphangion can be principally described as a short-distance pump whose primary "task" is to drive a bolus of fluid down the next one or few lymphangions. But together, the chain of such pumps is able to maintain effective long-distance transport of lymph. During the active pumping of lymphangions, coordinated contractions of the lymphatic muscle cells create an increase in intralymphatic pressure and form a local positive pressure gradient near the downstream front of the propagating contracting zone to propel lymph centripetally. At the upstream edge of the contracting zone, a negative pressure gradient transiently occurs between contracting lymphangions and upstream relaxing lymphangions. This gradient generates the short-lasting localized reversed flow that closes the valve (McHale and Roddie 1976; Gashev 1991). Therefore the phasic contractions of lymphangions generate flow and shear in the lymphatics by themselves. Recently (Gasheva et al. 2006) we discovered that lymphangions have self-regulatory control mechanisms to maintain the efficiency of their diastolic filling and subsequent contractions. The shear that is generated during the active contractions of lymphangions cause fast nitric oxide (NO) release to promote easier lymphatic filling and increase the efficiency of the lymphatic phasic contractions; by this mechanism the flow/shear-dependent self-regulatory mechanisms in the lymphangions continuously adjust the lymphatic tone and phasic contractions to the physiologically variable preloads and outflow resistances (Gasheva et al. 2006). Intrinsic contractions of the lymphangions are initiated by cardiac-like action potentials (Van Helden 1993; Allen et al. 1986; Orlov and Lobov 1984; Mislin 1971). We and others have demonstrated that these contractions are modulated by hydrodynamic factors as well as different vasoactive substances and pharmacological agents (Benoit and Zawieja 1994; Gashev and Zawieja 2001; von der Weid 2001; Gashev 2002; Bridenbaugh et al. 2003; Zawieja et al. 1993, 1999; Benoit et al. 1989; Zawieja and Davis 1993). Furthermore our recent studies demonstrated regional differences in the active lymph pumps (Muthuchamy et al. 2003; Gashev et al. 2004). However, the mechanisms that regulate lymph transport are not completely established. A number of important aspects of lymphatic function are almost completely unknown including how the age-related impairment of lymph flow develops in lymphatic beds, what are the changes in the contractile and regulatory proteins in lymphatic muscle cells and endothelial cells and how they lead to regression of the functional adaptive reserves in lymphatic vessels due to the impaired pressure-dependent contractile/pumping activity of the lymphatic muscle cells and their endothelium/flow-dependent modulation. While the general influence of aging on several body systems has been studied, the impact of aging on the lymph transport systems that serve these body systems remain to be elucidated. Thus, addressing all of the above raised questions is important to obtain basic knowledge that will improve any attempts to treat age-related organ dysfunctions.

Dysfunctional lymphatic transport can result in a wide range of disturbances including edema, altered immune cell circulation, depressed immune function, impaired lipid metabolism, etc. The lack of knowledge on how aging affects lymphatic vessels certainly contributes to the situation when researchers and clinicians are faced to ignore lymph transport-related components of various diseases. Thus, investigations of the mechanisms affecting lymphatic contractile function during aging is extremely important for ongoing attempts to better understand lymphatic system and to discover the pathogenesis and the effective treatment of various aging-associated disorders.

Until recently there were no published reports on systematic studies on aging-associated changes in the active lymph pumps. However, a study was published more than 2 decades ago on the morphological structure of the lymphatic vessel walls of various lymphatic vessels during aging in humans (Orlov et al. 1983). The authors showed (Orlov et al. 1983) that in older human, the destruction of the elastic elements and atrophy of muscle cells in the thoracic duct wall resulted in the development of "duct sclerosis." Investigations of the human mesenteric lymphatic bed demonstrated that after 65, the number of collecting lymphatic vessels in the human mesentery was significantly reduced, and the number of connections between lymphatic vessels of the same lymphatic arcade level was greatly diminished. Moreover, the authors observed considerable signs of degradation of the blood capillaries in the lymphatic wall after the age of 60. In some preparations of collecting lymphatic vessels, aneurism-like formations containing only endothelial cells in their walls were found, primarily in the areas located downstream but close to the lymphatic valves. Due to the profound difficulty of measuring lymph flow in vivo, there are only a few reports demonstrating the measurements of reduced lymph flow in aged animals (Bulekbaeva 1988; Hollander and Dadufalza 1990; Chevalier et al. 1996). In particular, it was reported (Chevalier et al. 1996) that aging significantly decreased lymph flow from the main mesenteric lymph duct in rats by ~60 % between ages of 3 and 22 months. Recently, we obtained important functional and molecular evidences of the aging-associated alterations of contractility in thoracic duct and mesenteric lymphatic vessels (MLV), which we will summarize below.

Analyzing the contractile activity of the isolated aged rat thoracic duct segments and comparing these data with those obtained from their adult counterparts we found that transmural pressure-dependent regulation was altered in aged thoracic duct especially at higher levels of pressure. The imposed flow-dependent pump inhibition was completely abolished in aged thoracic duct, yet nitric oxide synthases (NOSs) blockade by Nω-Nitro-L-arginine methyl ester hydrochloride (L-NAME) (10^{-4} M) increased pumping in a flow-independent manner. Western blot analyses indicated that the relative levels of endothelial nitric oxide synthase (eNOS) were decreased in the 24-month-old thoracic duct when compared with 9-month-old thoracic duct, whereas iNOS levels were increased in 24-month-old thoracic duct (Gasheva et al. 2007).

In particular in these experiments we found that lymphatic tone index in the control group thoracic duct (TD) segments isolated from 9-month-old Fischer-344 (F-344) rats was 5–6 % at all levels of transmural pressure, which is in the same range of values for the tone in TD from our previous study performed with adult Sprague–Dawley rats (Gasheva et al. 2006). In the TD segments isolated from 24-month-old rats lymphatic tone index was in average 31 % lower than in 9-month-old segments. Furthermore, contraction amplitude was reduced in 24-month-old segments in comparison with segments from adult animals indicating the age-related negative inotropy in TD. The contraction amplitude decreases were 35, 44, and 64 % below contraction amplitude in 9-month-old control at 1, 3, and 5 cm H_2O transmural pressures, respectively (In addition, aging caused a negative chronotropy in TD—the reduction of *contraction frequency* in TD segments from 24-month-old rats). TD segments from 24-month-old exhibited 27, 34, and 41 % less contraction frequencies when compared with that of TD segments from 9-month-old rats at 1, 3, and 5 cm H_2O transmural pressures, respectively. As a result of the age-related negative inotropy and chronotropy in the TD, its pumping ability was greatly reduced in a 24-month-old group. Furthermore, fractional pump flow in aged thoracic segments was 1.9, 1.9, and 3.2-fold down compared to their adult (9-month-old) counterparts at 1, 3, and 5 cm H_2O transmural pressures, respectively.

We have recently demonstrated that spontaneous phasic contractions in TD produce an NO-dependent decrease in lymphatic tone that maintains the efficiency of the lymphatic contraction and subsequent lymphatic diastolic filling (Gasheva et al. 2006). In this discussed study we investigated how aging influences this self-regulatory mechanism in TD. All parameters of active lymph pump are modulated in the presence of L-NAME in TD from 9-month-old F-344 rats. These data are identical to the results of studies performed on TD segments isolated from adult Sprague–Dawley rats (Gasheva et al. 2006), indicating the NO pathway is operative in the 9-month-old F-344 rats. However, we found that this mechanism was altered in 24-month-old F-344 rats. Although in the presence of L-NAME, the lymphatic tone index was increased to the same degree in 9- and 24-month-old rats; the contraction amplitude and the contraction frequency were unaffected after administration of L-NAME in TD segments isolated from 24-month-old animals.

We then compared the imposed flow-induced contractile responses in TD segments isolated from 9-month- ($n=9$) and 24-month- ($n=7$) old rats (in different sets of experiments) before and after abluminal administration of L-NAME at 10^{-4} M. In TD segments taken from 9-month-old F-344 rats, the typically imposed flow-induced inhibition of lymphatic contractility was seen with the same trends of changes in the parameters of active lymph pump in TD we described in our previous studies with Sprague–Dawley rats (Gashev 2002; Gashev et al. 2002, 2004, 2006). During the increases in imposed flow up to 5 cm H_2O, lymphatic tone index decreased ~20 %. In addition, significant negative inotropic (decrease in contraction amplitude) and negative chronotropic (decrease in contraction frequency) effects were noted as we increased imposed flow in these TD segments. As a result of such influences of the imposed flow, the fractional pump flow was greatly diminished in 9-month-old segments of TD especially at high levels of imposed flow through it. Particularly at an imposed flow of 5 cm H_2O, the fractional pump flow was 4.1 times lower than at the conditions with no imposed flow, suggesting the presence of an imposed flow-induced inhibition of the active lymph pump in TD segments obtained from adult 9-month-old animals. The presence of L-NAME completely abolished this imposed flow-induced inhibition of the active pump parameters. All parameters of lymph pump in adult segments remained unchanged during the increases in the imposed flow after NOS blockade induced by L-NAME application.

On the contrary, the imposed flow-dependent sensitivity was completely absent in TD segments obtained from aged 24-month-old animals. The lymphatic tone index and contraction frequency parameters were not changed during imposed flow elevation in aged TD segments; contraction amplitude was slightly but not significantly reduced when imposed flow was set on 1 cm H_2O but did not change with further increases of imposed flow. As a result, the fractional pump flow remained unchanged in 24-month-old specimens during the imposed flow elevations, indicating that imposed flow-dependent sensitivity was completely abolished in aged TD. Furthermore, we did not observe the imposed flow-induced inhibition in 24-month-old TD segments after administration of L-NAME. The application of L-NAME induced the increases in lymphatic tone index, contraction frequency and fractional pump flow, and slightly decreased the contraction amplitude in these vessels; however, the imposed flow did not have any significant effects on these parameters.

To further determine whether NOS expression in the TD vessels from aged animals would have been altered, we performed western blot analyses for the proteins from 9- and 24-month-old rats. The relative levels of eNOS are not detectable or significantly decreased in the protein samples from the 24-month-old rat when compared to that of from 9-month-old animals. We found an increase in iNOS protein levels in the TD from 24-month-old rats comparing to the samples from an adult animal. Quantitative analyses from each group show a significant decrease (0.14 ± 0.04) in the relative level of eNOS and a significant increase in the relative level of iNOS (10.65 ± 1.30) in the TD segments from 24-month-old rats.

These data demonstrate age-related alterations of active pump in rat TD. The transmural pressure/stretch-dependent modulation of lymphatic contractility is one of the principal regulatory mechanisms of lymphatic pumping which serves a goal to adapt lymphatic transport to the different lymphatic preloads (Gashev 1989, 2002; McHale and Roddie 1976; Benoit et al. 1989; Gashev et al. 2004; Ohhashi et al. 1980). The alterations in stretch-related regulatory mechanisms in 24-month-old segments indicate that both pacemaking and contractile machinery are involved in age-related changes of active lymph pump in rat TD. Lowered lymphatic tone in aged vessel segments together with decreased lymphatic contraction amplitude may be considered an indicator of age-related weakening of muscle cells and their diminished ability to create enough force to maintain the level of tone and contractile force appropriate to the lymphatic preloads: at a comparatively low pressure level of 1 cm H_2O, the contraction amplitude is moderately lower in aged segments of TD. At higher pressure levels, the contraction amplitude is diminished in greater degree in aged TD and reached the statistically significant difference levels between 9- and 24-month-old specimens. This negative age-related inotropy in thoracic duct was accompanied with alteration in function of lymphatic pacemaking: frequency of lymphatic contraction was diminished, especially at high levels of transmural pressure (5 cm H_2O). Such negative age-related chronotropy in TD together with negative inotropy led to a greatly decreased functional pump flow indicating the diminished pumping ability of aged TD. The differences between stretch-induced responses in adult and aged animals are greater at higher levels of transmural pressure, suggesting also a diminished ability of aged TD to adapt its contractility to increased preloads. Thus, the aging of the TD leads to decreases in its functional reserves to adapt the contractility and pumping to the increased levels of lymph inflow in it. At higher levels of preload in the aged TD, its possibility to serve the increased demand in pumping through the duct will be diminished. Consequently, partial or complete failure to provide the adequate transport of lymph through the duct may occur.

Aging also alters the contraction-initiated self-regulatory mechanism in TD that serves to maintain lymphatic contractions in efficient energy-saved mode (Gasheva et al. 2006). Even though administration of L-NAME in aged thoracic duct caused increases of lymphatic tone to the same values observed in adult specimens, the contraction amplitude and contraction frequency did not demonstrate the regular patterns of changes after NOS blockade as we observed for 9-month-old segments in F-344 rats or as we have previously shown in TD segments obtained from adult Sprague–Dawley rats (Gasheva et al. 2006). Such alterations in the contraction-initiated self-maintained regulation in TD may also have an impact on its pumping ability and responsible for weakening of its contractile reserves. The fact that fractional pump flow (pumping) was slightly increased in aged thoracic duct in the presence of L-NAME may indicate the presence of inhibitory influences of the constant release of nitric oxide in the aged thoracic duct independent of the phasic contractions.

The findings described above correlate well with the data obtained in experiments with increased imposed flow in thoracic duct. In these experiments the 9-month-old segments demonstrated the regular (Gashev 2002; Gashev et al. 2002, 2004, 2006) pattern of imposed flow-induced inhibition. Administration of L-NAME completely abolished this inhibition. Fractional pump flow in 9-month-old specimens remained unchanged even at high levels of imposed flow after NOS blockade. Several studies have shown the involvement of the lymphatic endothelium in the NO-mediated modulation of lymphatic contractile activity (see most recent reviews in (Gasheva et al. 2006; Ohhashi et al. 2005; Gashev 2008; Gashev 2010)). Particularly, specific immunostaining of eNOS was demonstrated in the endothelial cells of bovine lymphatics (Marchetti et al. 1997) or in rat lymphatic endothelial cells (Hayes et al. 2003). In functional experiments it was demonstrated that precontracted dog thoracic duct rings relaxed to acetylcholine (ACh) or sodium nitroprusside (SNP) (Ohhashi and Takahashi 1991). Removal of the endothelium blocked the ACh-induced relaxations, which were suppressed or abolished by pretreatment with oxyhemoglobin, methylene blue, and L-NMMA. In other studies, ACh produced negative chronotropic and inotropic effects on the spontaneous contractions in isolated bovine mesenteric lymphatics (Yokoyama and Ohhashi 1993) that were dependent on the endothelium. Furthermore L-NMMA suppressed these responses, but did not eliminate them. In addition, it had been demonstrated that L-arginine completely reversed the inhibition of the ACh-induced responses by L-NMMA. The sources of NO involved in the regulation of lymphatic pump activity of rat mesentery were studied in situ (Shirasawa et al. 2000). L-NAME (nonspecific NOS-blocker) caused increases in contraction frequency and pump flow in lymphatics, and decreases in the lymphatic diameters. L-arginine reversed these effects while aminoguanidine, the iNOS inhibitor, had no effect. The authors conclude that endogenous NO-inhibited lymphatic pump activity and that these effects were mediated by eNOS. Our recent studies demonstrated the involvement of NO on imposed flow-induced inhibition in rat MLV (Gashev et al. 2002). Taken together all literature data described above and our data obtained in this study, we concluded that in 9-month-old segments of rat TD the imposed flow-induced inhibition of the active lymph pump was completely dependent on the NO released due to the activity of eNOS.

Interestingly, TD segments from 24-month-old rats behaved differently during the increases in imposed flow. The TD segments from aged group did not exhibit any significant imposed flow-dependent inhibition of the parameters of the active lymph pump, which remained unchanged during the imposed flow elevations. We concluded that eNOS-mediated imposed flow-dependent regulatory mechanism was completely depleted in 24-month-old TD segments. Therefore the ability of aged TD to adapt its pumping to the different levels of extrinsic lymph flow was severely altered. Moreover, the administration of L-NAME moderately increased lymphatic pumping in 24-month-old TD segments independent from the value of imposed flow. These findings indicated the presence of a flow-independent but NO-dependent inhibition of lymphatic contractions in aged TD. We propose that this inhibition exists in aged lymphatic segments due to the activation of iNOS. To confirm the functional data on pressure- and flow-independent NO-dependent inhibition of the active lymph pump in aged TD, we performed western blot analyses of eNOS and iNOS in samples isolated from 9- and 24-month-old TD. Data obtained in these experiments clearly indicated that the relative levels of eNOS were decreased in the 24-month-old thoracic duct when compared with that of 9-month-old thoracic duct, whereas iNOS levels were dramatically increased in 24-month-old thoracic duct. The detailed functional tests of aged TD described above clearly demonstrated that it is too simplistic to attribute the diminished contractility only to the sclerosis of aged TD (Rabinovitz and Saphir 1965) and/or to atrophy of muscle cells in its wall (Orlov et al. 1983; Rabinovitz and Saphir 1965).

In next the study we evaluated the aging-associated changes in pumping of MLV in adult (9-month-old) and aged (24-month-old) F-344 rats (Nagai et al. 2011). These data demonstrated a severe weakening of the lymphatic pump in aged MLV including diminished lymphatic contraction amplitude, contraction frequency and as a result—lymphatic pump activity. The data also suggest that the imposed flow gradient-generated shear-dependent relaxation does not exist in aged rat MLV, and the sensitivity of both adult and aged MLV to such shear cannot be eliminated by NOS blockade. These data provide new evidence of lymphatic regional heterogeneity for both adult and aged MLV. In MLV, a constant interplay between the tonic and phasic components of the myogenic response and the shear-dependent release of nitric oxide predominantly determines the level of contractile activity; the existence of another shear-dependent but NO-independent regulatory mechanism is likely present. Aging remarkably weakens MLV contractility, which would predispose this lymphatic network to lower total lymph flow in resting conditions and limit it's ability to respond to an edemagenic challenge in the elderly.

In this study(Nagai et al. 2011), we performed for the first time a detailed evaluation of the parameters of the active lymph pump in MLV isolated from 9- and 24-month-old F-344 rats. In the first part of this study, we compared the pressure/stretch-induced changes in contractility of isolated segments of MLV obtained from 9- and 24-month-old F-344 rats. Additionally, we used NOS blockade by 100 μM L-NAME to evaluate the importance of phasic contraction-generated, shear-dependent NO release by MLV in modulating these contractile responses in vessels obtained from both age groups. In the 9-month group, we observed that the lymphatic tone was diminished 32 % during the increases in transmural pressure from 1 to 7 cm H_2O, and the contraction amplitude changed in a bell-shaped fashion being highest at a transmural pressure of 3 cm H_2O with decline at higher pressures. An increase in vessel stretch induced a positive chronotropic effect leading to a 96 % increase in contraction frequency of MLV segments as the transmural pressure was raised from 1 to 7 cm H_2O. Cumulatively, maximum lymphatic pumping was observed at 3 cm H_2O. In general, MLV segments obtained from 9-month-old (adult) animals demonstrated a classic inotropic response to increases in wall stretch as previously described for these (Gashev et al. 2002, 2004, 2006) and other lymphatic vessels (McHale and Roddie 1976; Gashev 1989; Ohhashi et al. 1980).

Recent studies performed in vivo (Bohlen et al. 2009) demonstrated the presence of phasic NO release in rat MLV. While such phasic contraction-generated shear-dependent release of NO was predicted and pharmacologically confirmed in isolated TD experiments (Gasheva et al. 2006), their role in the regulation of isolated MLV contractions has never been evaluated. In our present experiments utilizing MLV segments obtained from 9-month-old rats, we found that NOS blockade induced remarkable increases in lymphatic tone in MLV quite similar to the observations made in isolated TD segments (Gasheva et al. 2006). After L-NAME administration, lymphatic tone increased significantly at all levels of transmural pressure (157 and 83 % compared to control at 1 cm H_2O and 7 cm H_2O, respectively). At the same time, treatment of isolated MLV segments with L-NAME did not alter contraction amplitude but did increase contraction frequency. After L-NAME application, contraction frequency was significantly higher at pressures 1, 3, and 5 cm H_2O (99 %, 57 %, and 57 %, respectively). As a result of the observed changes in vessel chronotropy, their minute productivity (indicated by FPF) was significantly greater after L-NAME application at transmural pressures of 1, 3, and 5 cm H_2O (52 %, 63 %, and 74 %, respectively).

In the 24-month group, segments of MLV under control conditions also demonstrated stretch-dependency of their contractile behavior. Directions of changes in lymphatic contractile parameters in the aged group were similar to those observed in the adult group including decreased lymphatic tone, bell-shaped changes in contraction amplitude, and positive chronotropy during increases in transmural pressure from 1 to 7 cm H_2O with maximums of lymphatic pumping observed between 3 and 5 cm H_2O. However, when comparing lymphatic contractions between 9-month and 24-month-old MLV, we noted consistent evidence of aging-associated contractile decay. While lymphatic tone was unchanged, both the amplitude and frequency of lymphatic contractions were significantly reduced in aged MLV. Aging-associated alterations in lymphatic contractility resulted in a 28–54 % lowering of the contraction amplitude in aged MLV compared to adult at different levels of transmural pressure. The negative chronotropic effects of aging on MLV contractility varied from 81 % depletion at 1 cm H_2O to 53 % depletion at 7 cm H_2O compared to adult MLV. Because of the aging-associated changes in MLV contractility described above, the minute productivity of aged vessels was significantly depleted 6.6-, 3.4-, 2.7-, and 3.2-fold at the transmural pressures of 1, 3, 5, and 7 cm H_2O, compared to corresponding levels of transmural pressure in adult MLV, respectively.

The results of the experiments utilizing L-NAME application in aged MLV segments demonstrated additional differences between adult and aged groups during the increases in transmural pressure. Surprisingly, we found that NOS blockade was unable to increase lymphatic tone in the aged group. No differences in tone were observed between control and L-NAME treatment conditions in 24-month-old MLV while 9-month-old MLV displayed significantly higher tone following

L-NAME treatment compared to control conditions and to conditions after L-NAME treatment in 24-month-old MLV.

Contraction amplitude was increased in aged vessels after NOS blockade with a significant 100 % increase observed at 1 cm H_2O transmural pressure. With further elevations of transmural pressure, this post-L-NAME positive inotropy progressively declined to insignificant levels (33 % increase at 3 cm H_2O transmural pressure to 9 % increase at 7 cm H_2O transmural pressure). Interestingly, L-NAME application in aged MLV restored the contraction amplitude to the values in adult control vessels at all levels of transmural pressure except 7 cm H_2O at which the contraction amplitude in aged L-NAME-treated MLV was significantly lower than in adult non-treated vessels. Frequency of lymphatic contractions in aged vessels was only slightly increased after L-NAME administration compared to control conditions at all levels of transmural pressure and these differences did not reach statistical significance.

As a result of the above post-L-NAME changes, the minute productivity of aged MLV was partially restored during NOS blockade. FPF of L-NAME-treated, aged MLV segments at 1 cm H_2O was significantly increased (3.4-fold) compared to aged MLV under control conditions but was still significantly less (48 %) than FPF in adult MLV during the control conditions. L-NAME treatment tended to increase FPF in aged MLV at all other levels of transmural pressure, but this trend increase became less dramatic and nonsignificant statistically at higher levels of transmural pressure (81 % at pressure 3 cm H_2O, 24 % at pressure 5 cm H_2O, and 47 % at 7 cm H_2O). FPF in L-NAME-treated aged MLV was always significantly lower than FPF in adult vessels during the control part of the tests at these pressure levels.

In the second part of this study (Nagai et al. 2011), we compared the imposed flow gradient-generated shear-dependent changes in contractility of isolated segments of MLV obtained from 9- and 24-month-old F-344 rats. In addition, we used NOS blockade by L-NAME (100 μM) to evaluate the importance of NO release by MLV in mediating these contractile responses in vessels obtained from both age groups. In the 9-month group, we observed that during the increases in imposed flow gradient from 0 to 5 cm H_2O, the lymphatic tone, contraction amplitude, and frequency diminished 19 %, 18 %, and 25 %, respectively, over this range of imposed flow gradients. Because of the observed changes, lymphatic pumping (indexed by FPF) declined (30 %) in the presence of the highest selected imposed flow gradient of 5 cm H_2O (compared to 0 cm H_2O of imposed flow gradient). In general, MLV segments from the 9-month-old (adult) animals under control conditions demonstrated the same type of the well-described (Gashev et al. 2002, 2004, 2006; Gashev 2010) contractile responses to increases in steady wall shear stress, induced by raise of the imposed flow gradient applied to vessels.

In our current study on isolated MLV segments obtained from 9-month-old rats, we found that NOS blockade induced a remarkable increase in lymphatic tone in MLV quite similar to the results observed in MLV segments obtained from SD rats (Gashev et al. 2002). After L-NAME administration in adult MLV, lymphatic tone significantly increased 64–73 % at the different levels of imposed flow gradient compared to control (no L-NAME treatment) conditions. Interestingly, after

L-NAME treatment of 9-month-old MLV, we still observed an 18 % decrease in lymphatic tone at 5 cm H_2O imposed flow gradient compared to no imposed flow gradient.

The L-NAME treatment of isolated adult MLV segments has no effect on the contraction amplitude. We coincidentally noticed that after NOS inhibition, the pattern of the imposed flow gradient-generated shear-induced weakening of the lymphatic pump was similar to that observed in the absence of such treatment. After L-NAME treatment during the increases in imposed flow gradient from 0 to 5 cm H_2O, the contraction amplitude diminished 16 %. We also found that NOS blockade significantly increased contraction frequency in the 9-month-old MLV segments 63–88 % at the various levels of imposed flow gradient compared to control (no L-NAME treatment) conditions. Interestingly, after L-NAME treatment of 9-month-old MLV, we still observed a 14 % decrease in contraction frequency at 5 cm H_2O imposed flow gradient compared to no imposed flow gradient. Because of the observed changes in the 9-month-old MLV contractile characteristics after L-NAME application, the minute productivity of these vessels (indexed by FPF) was significantly greater (57–80 %) at all levels of imposed flow gradient. NOS blockade was not able to eliminate the 25 % reduction in lymphatic pumping (FPF) induced by the highest selected imposed flow gradient of 5 cm H_2O (compared to the absence of the imposed flow gradient).

In the 24-month-old segments of MLV under control conditions, we found that during the increases in imposed flow gradient from 0 to 5 cm H_2O, the lymphatic tone, contraction amplitude, and contraction frequency diminished 19 %, 40 %, and 36 %, respectively. Because of these observed changes, lymphatic pumping declined 30 % in the presence of the highest selected imposed flow gradient of 5 cm H_2O (compared to no imposed flow gradient). In general, MLV segments from the 24-month-old (aged) animals under control conditions demonstrated contractile responses similar in magnitude and polarity to the changes observed in 9-month group under control conditions during the same increases in imposed flow gradients.

Following L-NAME application, we found differences between adult and aged MLV segments in response to increasing imposed flow gradients. In these experiments as well as in the pressure/stretch-related experiments described earlier, we discovered that NOS blockade was unable to increase significantly lymphatic tone in the aged group. Over the range of imposed flow gradients (from 0 to 5 cm H_2O), lymphatic tone decreased 19 % in the aged L-NAME-treated MLV. Contraction amplitude of aged vessels after NOS blockade was increased slightly but nonsignificantly. At 5 cm H_2O of imposed flow gradient, contraction amplitude in aged post-L-NAME MLV was still 22 % lower than that measured under no imposed flow gradient. Compared to no imposed flow conditions, the frequency of lymphatic contractions in aged MLV was insignificantly decreased; however, this subtle evidence of imposed flow gradient-dependency of the contraction frequency in aged MLV was eliminated after NOS blockade. Finally, aged MLV minute productivity only tended to increase in the absence of NO at all levels of imposed flow gradients; however, FPF in L-NAME-treated aged MLV was always significantly lower than

FPF in adult vessels during the control part of the tests at these levels of the imposed flow gradients.

In discussed study (Nagai et al. 2011), we completed the first comprehensive evaluation of aging-associated alterations in the contractile responses of MLV. Our new data allow us to make several important conclusions that widen our basic knowledge of the regulatory mechanisms of mesenteric lymph flow in adulthood and senility, demonstrating therefore the complex nature of lymphatic system. Our previous studies have demonstrated regional heterogeneity in the contractile behavior of lymphatic vessels (Muthuchamy et al. 2003; Gashev et al. 2004, 2006). Our current studies provided the first evidence of regionally different influence of aging on different lymphatic networks. To illustrate such conclusions, we compared isolated MLV data with published data (Gasheva et al. 2007) obtained from experiments utilizing isolated rat thoracic duct segments obtained from similar groups of adult (9-month-old) and aged (24-month-old) F-344 rats and treated with similar experimental conditions.

Thus, we performed a series of tests to determine the contractile responses of MLV isolated from 9-month-old (adult) animals to increase in wall stretch generated by different levels of transmural pressure. These tests provided adult control data for comparison to data obtained with aged MLV, and expanded our knowledge on the basic contractile characteristics of the adult MLV highlighting their important differences with the lymphatic contractions in the thoracic duct. In the first set of the experiments, we blocked the NOS in conditions without steady flow through the vessel segment, induced by the imposed flow. In this set of experiments, the endothelial cells in the MLV therefore only experienced shear stress due to the flow generated by the spontaneous phasic contractions of the lymphatic muscle cells. No exogenous pressure gradient favorable to flow was present in the lymphatic specimen. This contraction-generated "active" flow in MLV is responsible for phasic fluctuations in wall shear stress (Dixon et al. 2006) and leads to the appearance of spikes of NO release by the lymphatic endothelium that are synchronous with lymphatic contractions (Bohlen et al. 2009). By analyzing the physiological importance of such phasic contraction-generated shear-dependent regulation in thoracic duct (Gasheva et al. 2006), we previously made following important conclusions. This large lymphatic duct, with lower resistance to flow, does not need strong, long-distance lymph active pumping and presumably because this, and due to the low variability in lymph flow patterns during the different periods of day, the thoracic duct has a somewhat simple shear-dependent regulatory mechanism to support an energy-sufficient contractile pattern solely through NO-dependent self-regulation.

In the current study (Nagai et al. 2011) of adult MLV, we found that opposite to thoracic duct the NOS blockade by L-NAME cannot induce negative inotropy. The amplitude of lymphatic contractions remained the same after NOS blockade in adult MLV even though their contraction frequency increased over all selected levels of transmural pressure under such conditions. As a result, the minute productivity of the adult MLV increased after NOS blockade. It seems unlikely that, in natural in situ conditions, the NO molecules will serve a role to limit the productivity of the active lymph pump, and thus, the increase in FPF in isolated adult mesenteric vessels after L-NAME administration looks somewhat artificial. On the other hand, the fact that NOS blockade induced a significant increase in MLV tone and frequency (therefore limiting diastolic time and diastolic filling) allows us to expect by default that the contraction amplitude in the absence of NO should go down as it does in TD at the same conditions (Gasheva et al. 2006). Therefore, our data supports the assumption that the phasic shear-dependent regulation of MLV contraction interplays with additional important mechanism(s) to control the lymphatic contractile cycle in these vessels seems to be more complex compared to TD.

Our current data and data previously by us and others allow us to conclude that there is a complex interplay between the influences of phasic and tonic components of the stretch-dependent myogenic responses and the influences of phasic contraction-generated NO release in MLV. In the mesenteric lymphatic network, the lymph formation and lymph flow may increase very fast with digestion. During such periods, the intralymphatic pressure may rise dramatically fast resulting in the development of myogenic constriction (Davis et al. 2009), the tonic component of which, i.e., prolonged vessel constriction, by itself may lead to a local increase in vessel resistance to flow. The latter is unnecessary for lymphatic vessels during periods of high lymph formation. Stronger phasic contractions during the myogenic response ("compensatory increase in amplitude" (Davis et al. 2009)) will support stronger fluid propulsion, higher fluctuations in wall shear, and therefore higher phasic release of NO. The latter of which will enhance the diastolic relaxation effect through easing the lymphangion filling and diminishing subsequently local resistance to flow. In other words, after pressure elevation, the positive phasic inotropy during the development of the myogenic phasic response would counter balance the release of additional NO. In the rat thoracic duct where there is lower total resistance to flow, lesser necessity for strong active pumping, and generally smaller tone-induced changes in lymphatic diameter relative to maximal diameter, the tonic and phasic components of the myogenic response may play a less functional role. Conversely, the data indicates a dominant functional importance for nitric oxide release in the control of the contractility in this largest lymphatic vessel of the body (Gasheva et al. 2006; Gashev et al. 2004; Gashev 2008; Gashev and Zawieja 2010).

In our experiments with L-NAME, performed in (Nagai et al. 2011) we artificially misbalanced the cascade of contractile regulatory reactions by eliminating the phasic contraction-generated shear-dependent NO release. At the same time, we did not eliminate the phasic contraction-generated shear itself so the potential existence of an additional as of yet unidentified shear-dependent, but NO-independent, mechanism for the regulation of lymphatic contractile strength in the MLV cannot be excluded. Further support of this idea can be seen in results of our experiments with NOS blockade in MLV during periods of increased imposed flow. Opposite to adult thoracic duct, L-NAME cannot completely abolish the influences of the imposed flow gradient on the MLV contractile pump, which are similar to effects of LNMMA observed by us earlier in these lymphatic vessels (Gashev et al. 2002). In conclusion, these our findings in adult MLV under control and L-NAME-treated conditions provide new support for the complexity and regional variability in the mechanisms controlling lymph flow in the body, which still requires additional investigations.

Furthermore, when comparing the contractile activity of adult and aged MLV, we found that aging significantly diminishes their contraction amplitude (negative inotropic effect), and contraction frequency (negative chronotropic effect). Due to these aging-associated alterations in contractility, the minute productivity of aged MLV decreased 6.6-fold at lower levels of transmural pressure and ~3-fold at moderate and high levels of wall stretch in old vessels. Overall, we conclude that aging induces weakening of the mesenteric lymph pump potentially by altering both pacemaking and contractile mechanisms; the specific causes of such events still need to be discovered. We believe that the weakened aged MLV may not provide enough contractile force to facilitate the transport of newly formed, nutrient rich lymph from the gut thus elongating the duration of natural digestion-related events in whole body. Additionally, the aging-induced weakening of the mesenteric lymph pump may lead to a delay in the clearance of excessive fluid as well as potentially hazardous substances and foreign particles from the aged gut and mesentery. This negative impact of aging on the fluid/macromolecule balance in the gut occurs constantly in the elderly including the periods of functional rest when levels of lymph formation and lymphatic filling are relatively low. During such periods, aged MLV are experiencing lower levels of wall stretch, a point at which the aging-associated depletion of their pumping is most remarkable.

While analyzing the contractile behavior of aged MLV during transmural pressure elevations, we noted a greater negative chronotropic effect of aging in these vessels compared to the aged TD (Gasheva et al. 2007). As a result, the aging-associated depletion of MLV minute productivity at lower levels of transmural pressure is relatively greater than that observed in aged TD. Fractional pump flow in aged TD segments was 1.9- and 1.9-fold down compared to their adult (9-month-old) counterparts at 1- and 3-cm H_2O transmural pressures, respectively (Gasheva et al. 2007). In our current investigation of aged MLV segments, FPF was 6.6- and 3.4-fold down at the same levels of transmural pressure, respectively. The causes and functional importance of these differences in the ability of these two aged lymphatic vessels to work against increases in preloads (diastolic intraluminal pressure) are yet to be evaluated.

Experiments with the treatment of aged MLV by L-NAME in the absence of imposed flow demonstrated that NOS blockade influenced contraction amplitude in aged MLV while than it did not affect inotropy in adult vessels. The contraction amplitude is lower in aged MLV so the phasic contraction-generated shear-dependent release of NO in aged MLV presumably is also weaker than in adult MLV (smaller fluctuations in wall shear due to the weaker contractions). The fact that L-NAME application in the weakened aged vessels has a greater effect on inotropy than in adult vessels may indicate the existence of additional aging-associated shear-independent NO release in aged MLV. Future direct measurements of NO concentrations in aged MLV in situ will help provide clear understanding of these changes.

The objectives of the next study performed in our laboratory (Akl et al. 2011) were to evaluate the aging-associated changes, contractile characteristics of MLV, and lymph flow in vivo in male 9- and 24-month-old F-344 rats. Lymphatic diameter, contraction amplitude, contraction frequency and fractional pump flow, lymph flow velocity, wall shear stress, and minute active wall shear stress load were determined in MLV in vivo before and after L-NAME application at 100 μM. The active pumping of the aged rat MLV in vivo was found to be severely depleted, predominantly through the aging-associated decrease in lymphatic contractile frequency. Such changes correlate with enlargement of aged MLV, which experienced much lower minute active shear stress load than adult vessels. At the same time, pumping in aged MLV in vivo may be rapidly increased back to levels of adult vessels predominantly through the increase in contraction frequency induced by NO elimination. Findings support the idea that in aged tissues surrounding the aged MLV, the additional source of some yet un-linked lymphatic contraction-stimulatory metabolites is counterbalanced or blocked by NO release. The comparative analysis of the control data obtained from experiments with both adult and aged MLV in vivo and from isolated vessel-based studies clearly demonstrated that ex vivo-isolated lymphatic vessels exhibit identical *contractile* characteristics to lymphatic vessels in vivo.

With more details, in this study (Akl et al. 2011) we performed, for the first time, a detailed evaluation of the contractile activity of MLV and lymph flow in vivo in 9- and 24-month-old F-344 rats. In the first part of this study, we investigated, in detail, the contractile characteristics of MLV and lymph flow in adult (9-month) and aged animals (24-month) under control conditions. We found that in aged animals the lymphatic vessel diameters from the same location (group III by Benoit 1991) are significantly larger than in adult animals. The end-diastolic diameters and end-systolic diameters in 24-month-old MLV were 71 and 79 % greater than in their adult counterparts. At the same time, we observed only minor, nonsignificant lowering of the contraction amplitude in aged MLV vs. adult vessels under control conditions (20 and 25 % of diameter changes during the contractions, respectively). The aging-associated negative chronotropy was observed in all aged MLV: we noted threefold decrease in their contraction frequency compared to adult MLV. As a result of the described aging-associated changes in lymphatic contractile force (contraction amplitude) and pacemaking (contraction frequency), the minute active lymphatic pumping was significantly lower in aged animals. Both indices of the lymphatic pumping, AFP and FPF, were significantly diminished in the aged group with AFP 76 % lower (~4.2-fold lower) and FPF—77 % lower (~4.3-fold lower) than the adult group.

In addition, we analyzed the aging-associated differences in the characteristics of lymph flow in rat mesentery using selected single contraction cycles of "diastole-systole" with suitable contractile cycles during each experimental condition for the 9-month-old animals and for the 24-month-old animals. While the diastolic lymph flow velocity was slightly, but not significantly, higher in aged MLV, the maximal systolic lymph flow velocity was significantly (43 %) lower in aged animals. Correspondingly, we did not find any aging-associated changes in calculated diastolic (resting) wall shear stress but during the phasic contractions the lymphatic endothelial cells in aged MLV experienced an ~3-fold reduction of the maximal systolic wall shear stress.

We also compared the rate of change of the phasic contraction-generated (i.e., active) wall shear stress in both adult and aged groups and found a dramatic ~6.2-fold aging-associated decrease in 24-month-old MLV in comparison to the 9-month group. The phasic contraction-generated (active) minute wall shear stress "load" that the lymphatic endothelial cells experienced a minute, the ASFP, was ~9.7-fold lower in aged MLV.

Because of the importance of the NO molecule released by lymphatic endothelium for the regulation of lymphatic contractility and flow in adult (Gasheva et al. 2006; Ohhashi et al. 2005; Gashev 2010; Bohlen et al. 2009; Ohhashi and Yokoyama 1994; Mizuno et al. 1998; von der Weid et al. 2001) and aged (Gasheva et al. 2007; Nagai et al. 2011; Gashev and Zawieja 2010) lymphatic vessels, in discussed study (Akl et al. 2011) we implemented in vivo the local NOS blockade induced by topical administration of 100 μM of L-NAME. We compared the contractile behavior of MLV and lymph flow in adult and aged groups before and after the L-NAME administration. We found that the NOS blockade with a duration of 15 min induced slight, but not significant, constriction in both adult and aged MLV. Only end-systolic diameter in aged MLV was significantly decreased by 25 % after 15 min of the L-NAME application. During these small changes in lymphatic diameters, after the NOS blockade in MLV of both aged groups, the difference between the contraction amplitude in adult and aged MLV was reversed by L-NAME from slightly negative (20 % lower in aged group) to positive (50 % higher in aged group). The greatest observed influence of the NOS blockade was its chronotropic effect. While in adult MLV the contraction frequency was significantly increased (~2.1-fold) 15 min after the L-NAME application, in the aged MLV this increase in the lymphatic contraction frequency was over of 3.5-fold. The average contraction frequency of the aged MLV treated even only 5 min by L-NAME was 25 % (although not statistically significant) higher than the contraction frequency in adult lymphatic vessels under control conditions. The main paradox of the influence of the NOS blockade in aged MLV was found when analyzing the indices of their minute productivity, AFP and FPF, which increased in both age groups as consequence of chronotropic and inotropic influences of the L-NAME administration. In adult vessels, L-NAME application was able to increase significantly both of these indices by 57 % and 59 %, respectively. In aged MLV, the influence of the NOS blockade was remarkably greater. Specifically, AFP was increased by 538 %, and FPF by 511 % compared to the control conditions after 15 min of the L-NAME administration. Such influence of the elimination of NO on aged MLV was not only able to compensate the observed ~4.2-fold aging-associated depletion in minute productivity in the aged MLV lymph pump in vivo but, after only 5 min of the L-NAME application, was able to maintain this productivity at the same level as the L-NAME-treated adult lymphatic vessels.

Another important result for discussed study (Akl et al. 2011) was in our findings that the diastolic lymph flow velocity remained unchanged during the L-NAME application in both selected age groups. At the same time we observed a moderate, nonsignificant increase in the maximal systolic lymph flow velocity in the adult group, in aged MLV the maximal systolic lymph flow velocity was more than tripled after 15 min of the L-NAME treatment (~3.6-fold increase) thus being higher than in adult MLV under the same experimental conditions. Correspondingly, we did not find any aging-associated changes in calculated diastolic (resting) wall shear stress after the NOS blockade in both selected age groups while the maximal systolic wall shear stress was significantly increased in MLV of both selected ages (139 % increase in adult and 452 % increase in aged MLV). Although the absolute difference (roughly 7 dyn/cm^2) is similar, the percentage as a reflection of the basal control level of the maximal systolic wall shear stress does matter. By underlying the degree of changes, we are underlying here the general tendency of lymphatic pumping in aged MLV to be increased dramatically after NO elimination towards values observed in adult animals at the same conditions. Additionally, we compared effects of the L-NAME treatment on the rate of change of the phasic contraction-generated (i.e., active) wall shear stress in both adult and aged groups. We found its dramatic ~4.8-fold increase in 24-month L-NAME-treated MLV as compared to aged MLV under control conditions. Such changes after L-NAME administration nearly matched the rate of change in active wall shear stress for the NOS-blocked aged MLV vs. the NOS-blocked adult vessels, for which the rate did not change significantly from control during the L-NAME treatment. As a reflection of these changes, the phasic contraction-generated (active) minute wall shear stress "load", which the aged lymphatic endothelial cells experienced a minute after NOS blockade, showed the ASFP to increase by ~25.3-fold in comparison to aged MLV in control conditions. However, the ASFP did not change significantly over the NOS blockade in adult MLV during the same period of time.

Furthermore, in this study (Akl et al. 2011) we performed comparative data analysis of the contractile parameters of rat MLV obtained both in vivo and in isolated vessels experiments. We found that in control groups of 9- and 24-month-old MLV all three investigated contractile characteristics namely, contraction amplitude, contraction frequency, and fractional pump flow, obtained in isolated vessel-based experiments were not significantly different from those obtained in vivo. Moreover, after L-NAME administration, as described in this study for the in vivo experiments and previously (Nagai et al. 2011) for isolated vessels, all three parameters of contractile activity of the 9-month-old MLV were not significantly different between in vivo and isolated vessels data groups with the same degrees and directions of changes both in vivo and under isolated vessel conditions. The primary intriguing difference we found was in contractile chronotropy of the aged MLV in vivo after L-NAME administration as compared to the isolated vessel experiments with the same aged MLV under the same experimental conditions (NOS blockade). The contraction frequency of the 24-month-old MLV situated in vivo increased ~3.5-fold in comparison to the frequency of the same vessels under control conditions. As a result of this positive L-NAME-induced chronotropy and the 50 % (but nonsignificant) increase of the contraction amplitude, the minute pumping (FPF) of the L-NAME-treated aged MLV was ~6.1-fold higher than in the same aged vessels under control conditions. In other words, the NOS blockade of aged MLV in vivo was able to not only compensate the aging-associated deficiency of minute pumping in aged vessels under control conditions, but additionally enhance their minute pumping all the way up to the levels of NOS-treated adult 9-month-old MLV. In isolated aged MLV we observed much weaker, nonsignificant, positive chronotropy and a much smaller increase in pumping after L-NAME administration (Nagai et al. 2011).

Therefore we performed, for the first time, a detailed evaluation of the parameters of the contractility of aged MLV in vivo and characterized lymph flow in the aged mesenteric lymphatic network (Akl et al. 2011). Results were compared to the same characteristics of lymphatic contractility and flow in adult MLV. We performed these evaluations under control conditions and after an NOS blockade of 100 μM L-NAME. This allowed us to make important conclusions about the comparative roles of the NO-dependent regulatory mechanisms, which control the lymphatic contractility and lymph flow in vivo in the adult and aged body. Importantly, we performed, for the first time, the direct detailed comparison of the characteristics of the lymphatic contractility in vivo and contractile characteristics of the isolated lymphatic vessels exteriorized from the same location for both adult and aged MLV under both control conditions and after NOS blockade.

The usage of the adult animals for control group created additional opportunities to widen our basic knowledge of the normal lymphatic function when we performed a detailed characterization of the lymphatic contractile activity and lymph flow in the adult body, intact and under the conditions of disrupted synthesis of the NO molecule by topical L-NAME administration. Subsequent comparative analysis allowed us to make important conclusions on physiological relevance of the data obtained in isolated vessel-based experiments. We found that the characteristics of the contractile activity and lymph flow in MLV in vivo in 9-month-old F-344 rats under control conditions are similar to those reported in literature for young rats (Benoit et al. 1989; Dixon et al. 2006; Benoit 1991; Zweifach et al. 1972; Zweifach and Prather 1975; Brill et al. 2001; Galanzha et al. 2002). In particular, our data matched the contractile characteristics of the mesenteric lymphangions located closer to the intestine, described as group III in Benoit's study (Benoit 1991). As an example, lymphangions studied by Benoit (1991) contracted at resting conditions with a frequency of 7.3 ± 1.5 contractions/min, while in our current study—9.0 ± 0.6 contractions/min. Their average FPF, calculated from Table 2 in (Benoit 1991), was 4.16, while in our study (Akl et al. 2011) it was 3.9. Interestingly, in general, the same parameters of the contractility of MLV were found in young Sprague–Dawley rats (Benoit 1991) and adult F-344 rats (Akl et al. 2011). Such comparisons provide additional support to our earlier conclusions on the validity of the comparisons between adult F-344 and young Sprague–Dawley rats (Gasheva et al. 2007). These findings provide evidence that, in rats, the adult-like contractile activity of lymphatic vessels can be found in comparatively young ages, from 2 to 3 months, while later this activity stays in a plateau-like state to at least the early adult ages (9-month-old as in our studies).

Our analysis of the parameters of the mesenteric lymph flow in vivo in adult rats under control conditions demonstrated the same order of magnitude when compared to literature data available for the rat mesenteric lymphatic network. In particular, the values of average diastolic lymph flow velocity of 0.23 mm/s and maximal systolic lymph flow velocity of 1.81 mm/s correlated to those observed in young Sprague–Dawley rats reported in several studies (Dixon et al. 2006; Brill et al. 2001; Galanzha et al. 2002, 2005; Fedosov et al. 2002) and in young F-344 rats (Galanzha et al. 2005). The average values of diastolic and maximal systolic wall shear stress, 0.61 dyn/cm^2 and 5.15 dyn/cm^2, respectively, were also found to be similar to the ranges determined in young Sprague–Dawley rats (Dixon et al. 2006), which strengthen our general conclusions on the similar lymph contractility and flow patterns within young and adult rats of different strains. At the same time, we extended our analyses of lymphatic wall shear stress parameters and implemented, for the first time, the calculations of the rate of changes in active wall shear stress in MLV in vivo. Specifically, this parameter in average in adult animals was 16.23 dyn/cm$^2 \times s^{-1}$. Newly implemented estimations of the active wall shear stress-frequency product allowed subsequent comparisons of the relative minute active wall shear stress load in contracting lymphatic vessels under different conditions within selected age groups.

In addition to the comparisons with the literature data obtained on mesenteric lymph flow in vivo, we performed a comparative analysis of the main contractile characteristics of the MLV, namely contraction amplitude, contraction frequency, and fractional pump flow, obtained in isolated vessel-based experiments and in vivo experiments. As described in the Methods and Results sections, we pooled together the data from this study and re-normalized contraction amplitude data previously published from isolated vessel-based studies (Nagai et al. 2011). The characteristics of the contractile activity of the adult MLV under control "normal" conditions in vivo were not statistically different from those obtained during the isolated vessel-based experiments (done with the same vessels from similar animals under the same control conditions as in vivo). These data attract additional attention by the fact that, after performing the analysis, the averages of the contraction frequency and FPF in vivo were well positioned in the middle between the same parameters registered in isolated vessels at transmural pressures 1 and 3 cm H_2O. Therefore, it is reasonable to conclude that in our in vivo experiments the MLV were contracting at diastolic (resting) pressure of ~2 cm H_2O. This value matches the values for end-diastolic pressure of ~2 cm H_2O in MLV measured directly by the servo-null micropressure approach (Fig. 7 from Benoit et al. 1989). Moreover, comparative analysis of the contractile behavior of aged MLV (under control conditions) in vivo vs. the same type of isolated vessels demonstrated similar results for their aging-associated changes. The averages of contraction frequency and FPF in vivo also remained positioned between the same parameters registered in isolated vessels at transmural pressures 1 and 3 cm H_2O.

After performing this comparative analysis of the major contractile characteristics of the MLV in vivo and vessels isolated from the same location and from the same age/strain animals as those in vivo, we approached the important conclusion regarding methodology of the modern lymphatic research. Data analyses presented in this study, confirmed the fact that carefully isolated and properly maintained ex vivo, lymphatic vessels can demonstrate nearly identical characteristics of those in vivo in terms of their contractility and therefore reflect their normal physiological function. In light of our present data, the recent statements that ex vivo studies "may not recapitulate 'normal' functioning lymphatics" (Kwon and Sevick-Muraca 2007), being initially not supported by any scientific analysis, now can be finally excluded from further considerations.

On the other hand, a careful analytical comparison of the influences of the NOS blockade in the aged MLV in vivo and in isolated vessel-based studies allowed us to determine the important differences (discussed below) between functioning of the lymphatic vessels per se, and functioning of the same vessels under the additional influence of the aged tissue microenvironment.

As we mentioned above, there is little information available in literature dedicated to the aging-associated changes of lymphatic contractility and lymph flow. Therefore, the detailed evaluation of the status of lymphatic contractility and lymph flow in the aged body is an important initial task, which is able to provide additional, still mainly ignored knowledge on unknown but essential, elements of pathogenesis of many chronic disorders, which manifest or worsen with aging. While such evaluations of the aging of lymphatic functions are still in their infancy, we believe that the immediate discovery of aging-associated alterations of the regulatory mechanisms controlling lymph flow in the aged body is not possible before the completion of a careful descriptive characterization of the aging-altered parameters of lymphatic contractility and lymph flow started by this and similar studies.

The initial finding, which attracted our attention while comparing adult and aged MLV in vivo, was the fact that the aged lymphatic vessels from the same anatomical location within the mesenteric lymphatic network in rats of the same body weight have significantly greater resting lymphatic diameter. We consider this enlargement of the aged MLV as an indicator of the deep aging-associated remodeling of the lymphatic wall. With sparse literature data on sclerosis, enlargement, aneurism-like formations, muscle cell atrophy, and muscle layers' disorganization in aged lymphatic vessels (Orlov et al. 1983; Gashev 2010; Rabinovitz and Saphir 1965), our current findings (Akl et al. 2011) create a solid foundation for detailed follow-up studies on aging-associated alterations of the biomechanical properties of lymphatic wall.

Furthermore, we found a severe negative chronotropic effect of aging on the contractility of MLV in vivo. Specifically, their contraction frequency was depleted while contraction amplitude of the aged vessels was only slightly diminished. Consequently, we observed a profound aging-associated decrease in the minute productivity of the aged MLV in vivo that we mainly linked not to the altered degree of contractile displacement of the lymphatic wall, but to the diminished number of the contractile events in aged lymphatic vessels. At the same time, we believe that we discovered additional signs of sluggishness of the contractile events in aged MLV. Substantial reduction in the rate of change in the development of phasic contraction-generated wall shear stress indicates the slower development of systolic contractile force in aged lymphatic wall with slower generation of the lymph-propelling local axial pressure gradient. Greatly diminished maximal systolic lymph flow velocity may also be considered as an additional sign of slower and weaker generation of the pressure wave inside aged lymphatic vessels. This aging-associated decrease of the phasic contraction-generated (active) wall shear stress correlates with profound depletion in contraction frequency of the aged MLV. These two overlapping events of lymphatic aging cumulatively induced dramatic depletion in minute active wall shear stress load (~9.7-fold decrease in ASFP) to lymphatic endothelial cells, which is the first time has been described for aged lymphatic vessels. This aging-associated decrease in minute active wall shear stress load therefore creates a ground for disruption of the phasic contraction-generated shear/NO-dependent regulatory mechanisms of the lymphatic contractile events (Gasheva et al. 2006; Gashev 2010; Nagai et al. 2011; Bohlen et al. 2009; Gashev and Zawieja 2010) in aged mesenteric lymphatic network. Currently, there are not enough data to determine the nature of the aging-associated negative chronotropy in the MLV, existence of which was confirmed both in vivo as well as in isolated vessel-based experiments. We propose that the negative chronotropy and the consequent depletion of the minute active wall shear stress load may occur due to the yet unknown aging-associated alterations in lymphatic pacemaking. At the same time this decrease of minute active wall shear stress load will induce a reduction in the amount of phasically generated NO through depletion of already well-described mechanisms (Gasheva et al. 2006; Bohlen et al. 2009). Further, the diminished phasic NO release in aged MLV will consequently intensify the slowing of the lymphatic contractile events by reducing the rate of lymphatic diastolic relaxation and diastolic lymphatic filling (Gasheva et al. 2006), and therefore will slow the aged lymphatic contractility even more. We observed this in our current studies. However, to date the complex detailed evaluation of the functional importance for this newly discovered phenomenon of the aging-associated reduction in the minute active wall shear stress load in aged MLV remains to be performed. In addition, we want to underline here that the resting diastolic lymph flow velocity only elevated slightly while the corresponding resting diastolic wall shear stress remained unchanged (due to the vessel enlargement) in aged MLV.

Next, we implemented (Akl et al. 2011) topical application of 100 μM of the nonselective NOS inhibitor, L-NAME, in order to eliminate phasic contraction-generated and steady flow-generated release of NO in both adult and aged MLV. Such experimental conditions we created taking into account the importance of the NO molecule released by lymphatic endothelium for regulation of lymphatic contractility and flow in adult (Gasheva et al. 2006; Ohhashi et al. 2005; Gashev 2010; Bohlen et al. 2009; Ohhashi and Yokoyama 1994; Mizuno et al. 1998; von der Weid et al. 2001) and aged (Gasheva et al. 2007; Nagai et al. 2011; Gashev and Zawieja 2010) lymphatic vessels. It is crucial to underline here that in both adult and aged mesenteric tissues the topical application of L-NAME did not induce any generalized effects, like increases in lymph formation. This was confirmed by the absence of any changes in resting diastolic lymph flow velocity during the L-NAME administration. The importance of the exclusion of the nonspecific effects of the NOS blockade while investigating the contractility of MLV was illustrated by the report of Galanzha et al. (2002). In that study, the application of the nonselective NOS-blocker, L-NNA, induced constriction in half of the observed rat MLV, while the other half of the vessels demonstrated dilation. This may be considered a result of post-L-NNA increases in lymph formation in part of the animals and as a sign of corresponding increased volumetric load. Unfortunately, the separate evaluation of the diastolic and systolic lymph flow has not been done in that study (Galanzha et al. 2002) which creates difficulty for the interpretation of its inconsistent results.

At low levels of lymph flow (fasted rats), we found that the topical L-NAME administration, without increase in diastolic lymph flow, was able to induce changes in lymphatic tone, chronotropy and minute pumping of 9-month old MLV. However, constriction and increase of the contraction frequency during NO absence did not lead to decrease of the contraction amplitude as was found at low levels of flow for solely NO-modulated shear-dependent regulatory mechanisms of the thoracic duct contractility (Gasheva et al. 2006, 2007). Potentially, in the MLV another mechanism prevents NOS-blockade negative inotropy by maintaining contraction amplitude at the same level. This observation supports our previously expressed hypothesis on the potential existence of an additional yet unidentified shear-dependent, but NO-independent, mechanism for the regulation of lymphatic contractile strength in the MLV (Nagai et al. 2011).

An intriguing consequence of the topical L-NAME administration in aged MLV in vivo is the profound differences depicted in the chronotropic response between in vivo and isolated aged vessels. The contraction frequency of in vivo L-NAME-treated MLV was significantly greater than that during similar treatment in isolated vessels at transmural pressure 1 cm H_2O. Cumulatively the minute productivity of the aged MLV in vivo was significantly greater than that in isolated vessels at transmural pressures 1 and 3 cm H_2O during the NOS blockade. At the same time the L-NAME application in aged MLV in vivo was ont only able to compensate the aging-associated deficit in minute pumping but also to increase pumping in aged vessels up to levels of pumping in adult MLV. We conclude that MLV in adulthood possesses considerable contractile reserves such that even "chronically developed" alterations in muscle cell density and orientation in aged lymphatic wall (Gashev 2010) are not able to preclude the ability of the aged MLV to be rapidly (within 5 min) stimulated, under resting conditions, up to the levels of pumping in the adult MLV. At the same time the question remains to be answered, "Will the rapid L-NAME (or any other drug)-induced stimulation maintain the levels of aged minute lymphatic pumping similar to the adult counterparts during the periods of the increased volumetric load (lymph formation) in aged body, and, if so, how long will it be maintained?" Moreover, the observations that L-NAME administration in vivo induced ~2.5-fold greater increase of lymphatic pumping in comparison to isolated vessels moves us to the idea that in aged tissues surrounding the aged MLV, the additional source of some yet unidentified metabolites which stimulate lymphatic contractions and whose effect may be counterbalanced or blocked by NO release. The post-L-NAME observed increases in diastolic wall shear stress (due to the vessels constriction, but not the diastolic lymph flow velocity) and minute active wall shear stress load (due to the increases in lymphatic contraction frequency) in aged MLV cannot be linked to the described differences between in vivo and isolated vessels since the NOS function was blocked during both experimental conditions. As a matter of fact, we believe that a focus for future follow-up investigations on the nature of the aging-associated changes of pumping of the MLV to a large extent will relate to discovery of the mechanisms of potential interaction of the aged contractile lymphatic vessels and aged tissues surrounding them.

As a direct follow up of the studies with adult and aged MLV discussed above (Akl et al. 2011), we recently investigated the potential presence of permanent stimulatory influences in the tissue microenvironment surrounding the aged MLV which influence aged lymphatic function (Chatterjee and Gashev 2012). In this study, we performed immunohistochemical labeling of proteins known to be present in mast cells (mast cell tryptase, c-kit, prostaglandin D2 synthase, histidine decarboxylase, histamine, transmembrane protein 16A, and tumor necrosis factor-alpha) with double verification of mast cells in the same segment of rat mesentery containing MLV by labeling with Alexa Fluor 488 conjugated avidin followed by toluidine blue staining. Additionally, we evaluated the aging-associated changes in the number of mast cells located by MLV and in their functional status by inducing mast cell activation by various activators (substance P; anti-rat DNP Immunoglobulin E; peptidoglycan from *Staphylococcus aureus* and compound 48/80) in the presence of Ruthenium Red followed by subsequent staining by toluidine blue. We found that there was 27 % aging-associated increase in total number of mast cells, with ~400 % increase in number of activated mast cells in aged mesenteric tissue in resting conditions with diminished ability of mast cells to be newly activated in the presence of inflammatory or chemical stimuli. We conclude that higher degree of pre-activation of mast cells in aged mesenteric tissue is important for the development of aging-associated impairment of function of MLV. The limited number of intact aged mast cells located close to the mesenteric lymphatic compartments to react to the presence of acute stimuli may be considered contributory to the aging-associated deteriorations in immune response.

In more details, as we mentioned above as a result of our recent studies (Akl et al. 2011), we focused our new research efforts (Chatterjee and Gashev 2012) on understanding the nature of the aging-associated alterations of lymphatic functions in MLV through investigations of the mechanisms of potential interaction of the aged contractile lymphatic vessels and aged tissues surrounding them. One of the first candidates for us to investigate was mast cells situated in close proximity to MLV with their potential involvement into "lymphatic vessel—tissue" interactions in aged animals described above. Mast cells are long-lived tissue-resident cells which are strategically located at sites where host tissue comes in contact with external allergens and microbes such as upper dermis, respiratory tract, and bowel mucosa (Beil et al. 2000; Marshall 2004; Dawicki and Marshall 2007). Previous mast cell studies were limited to the context of allergy and asthma, but over the past decades mast cells have been extensively researched and have been found to play an important role in modulating innate immunity and in shaping the host adaptive immune response (Marshall 2004; Dawicki and Marshall 2007; Mekori and

Metcalfe 2000; Marshall et al. 2003). There are growing evidences suggesting that mast cells are also essentially involved in causing inflammatory changes during the course of chronic diseases: mast cells can produce, store, and release upon activation numerous inflammatory and vasoactive mediators that continuously modulate the immune system and promote inflammation (Malaviya and Abraham 2001; Malaviya and Georges 2002; Harvima and Nilsson 2011; Amin 2012). These mediators are categorized into preformed and newly synthesized mediators. Preformed mediators stored in the secretory granules of mast cells include proteases: tryptase, chymase, carboxypeptidase, and cathepsin G; biogenic amines like histamine and serotonin; proteoglycans like heparin sulfates and chondroitin sulphate E; acidic hydrolases; various cytokines and chemokines like IL-4, TNF alpha, IL-15, IL-6, IL-8, IL-13, monocyte chemotactic protein-1 (MCP-1); and vascular endothelial growth factors (VEGFs) (Amin 2012; Grutzkau et al. 1998; Boesiger et al. 1998; Galli et al. 2005; Detoraki et al. 2009; Lundequist and Pejler 2011). Newly synthesized mediators include prostaglandin D2, prostaglandin E2, leukotriene C4, leukotreine B4, platelet activating factor and varying amounts of other cytokines and growth factors like TNF alpha, IL-1, IL-3, IL-6, IL-9, IL-13, nerve growth factor (NGF), VEGFs, stem cell factor (SCF), granulocyte macrophage-colony stimulating factor (GM-CSF), and basic fibroblast growth factor (b-FGF) (Detoraki et al. 2009; Lundequist and Pejler 2011; Schwartz 1987; Theoharides et al. 2007). Mast cells can be induced to undergo de novo synthesis or upregulation of different cell surface molecules like chemokine receptors, toll-like receptors, MHC II molecules which promote an inflammatory reaction. Mast cell mediators like prostaglandin D2, leukotrienes B4, histamine, cytokines like TNF alpha, GM-CSF, IL-1, IL-3, IL-5, and chemokines like IL-8, regulated upon activation, normal t-cell expressed, and secreted (RANTES), MCP-1, eotaxin act as important chemoattractants for eosinophils, neutrophils, and dendritic cells which participate in subsequent development of the immune and inflammatory processes (Raible et al. 1992; Ramos et al. 2003; Shakoory et al. 2004; Salamon et al. 2005; Suto et al. 2006; Heib et al. 2007; Dawicki et al. 2010; Simon et al. 2011). Mast cells have been found to be situated in close proximity to blood vessels and sensory nerves and that many neuropeptides like substance P and corticotrophin-releasing hormone have been shown to activate mast cells (Mousli et al. 1989; Suzuki et al. 1995; Theoharides et al. 1998; Lorenz et al. 1998; Saban et al. 2002; Esposito et al. 2002; Lytinas et al. 2003). Close proximity of mast cells to blood vessels is important in hypersensitivity and inflammatory reactions where release of histamine from activated mast cells increases vascular permeability causing increased efflux of fluid and immune cells from the circulation to the interstitial tissue (Marshall 2004; Dawicki et al. 2010; Marshall and Jawdat 2004).

Although there has been extensive research on mast cells and the blood vasculature, few studies have investigated the modulatory role of mast cells on collecting lymphatic vessels. It has been known that mast cells are often situated in higher number in close proximity to the lymphatic vessel wall (Orlov and Lobov 1984; Kluth 1951; Wilhelm et al. 1978). In the paravascular zones of the mesentery, mast cells are particularly associated with lymphatic vessels, rather than with blood vessels (Yong et al. 1975, 1977). While mast cells produce and secrete various potent vasoactive mediators mentioned above, it was proposed by several investigators that at least some of them may influence lymphatic contractile/pumping function. In particular it has been shown that histamine is a potent dose-dependent modulator of lymphatic contractility (Orlov and Lobov 1984; Mislin 1971; Ohhashi et al. 1978; Johnston et al. 1983; Unthank and Hogan 1987; Ferguson et al. 1988; Watanabe et al. 1988; Dobbins et al. 1990; Fox and von der Weid 2002; Plaku and von der Weid 2006; Petunov et al. 2010; Pan'kova et al. 2011). Other mast cell-derived mediators like heparin (Orlov and Lobov 1984; Lobov and Pan'kova 2010), serotonin (Ohhashi et al. 1978; Johnston et al. 1983; Unthank and Hogan 1987; Dobbins 1998), leukotrienes B4 and C4 (Johnston et al. 1983; Ferguson et al. 1988), and thromboxane A2 (Plaku and von der Weid 2006; Sinzinger et al. 1984) are also shown to have physiological effects on lymphatic contractility. However until now, only in one study (Plaku and von der Weid 2006) the activation and degranulation of mast cells has been directly linked to subsequent changes in lymphatic contractility mediated by histamine. While it assumes that mast cell activation influences lymph contractility and flow, and should subsequently modulate immune reactions in the mesentery, lipid absorption, and interstitial fluid dynamics, there are no currently available data which directly establishes links between all of these processes. Moreover, the functional status of mast cells located in close proximity to aged lymphatic vessels has never been evaluated in comparison to that in adult body and therefore the aging-associated alterations in the regulatory interactions between lymphatic vessels and mast cells located nearby them remain undiscovered.

More detailed, in this discussed recent study (Chatterjee and Gashev 2012), we performed, for the first time, immunohistochemical labeling of protein targets shown to be present in mast cells with double verification of their presence in the same segment of rat mesentery containing MLV by labeling it with Alexa Fluor 488 conjugated avidin followed by staining of the same segment with toluidine blue. We found that mast cells are located in close proximity to rat MLV in segments of mesentery obtained from both age groups; however, it was impossible to detect any potential measureable differences between specimens from adult and aged rats using the immunohistochemical labeling technique. Next, for the first time, we evaluated the aging-associated changes in the number of mast cells located by MLV and in their functional status by using an experimental approach of mast cell activation induced by various activators and staining by Ruthenium Red, a commonly used dye to stain degranulated mast cells followed by staining of the same segment with toluidine blue. We discovered that the major difference between 9- and 24-month-old segments relays to the number of activated mast cells before experimental stimulation, which is significantly greater in 24-month-old animals. Using toluidine blue staining we were able to calculate the total number of mast cells in all mesenteric segments used in experiments with mast cells activation. We found a significant 27 % increase in the number of mast cells located by MLV in 24-month- vs. 9-month-old segments: 56 ± 3 and 44 ± 2 cells/segments respectively with $n = 84$ for 24-month-old specimens and $n = 73$ for 9-month-old specimens.

We determined that in all mesenteric segments in control conditions (before treatment) the number of activated mast cells in 24-month-old segments was in average 403 % higher than in 9-month-old segments before treatment. The number of mast cells activated in 24-month-old segments in untreated conditions for each type of treatment were statistically nonsignificant between the different types of treatment—namely, compound 48/80, substance P, PGN, IgE, and sham-treated samples. For activation by compound 48/80, the ratio of activated cells to total cells in 9-month-old specimens was increased 1,258 % after activation compared to control (before activation) conditions, while in the 24-month-old specimens we observed only 198 % increase. For Immunoglobulin E, Substance P, and peptidoglycan these numbers were 2,667/386 %, 818/280 %, and 318/130 %, respectively, while in sham control specimens routine incubation of the sample in PSS without adding mast cell activators did not induce any changes. To better illustrate the observed phenomena, we calculated parameter called as fraction of cells activated after treatment. We found that activation by compound 48/80 was able to activate in 9-month-old mesenteric segments 89.2 ± 1.4 % of available intact mast cells, while in 24-month-old segments—only 48.6 ± 4.8 %. For Immunoglobulin E, Substance P, and peptidoglycan, these numbers were $92.4 \pm 2.2/51.9 \pm 7.1$ %, $79.7 \pm 2.8/60.3 \pm 6.6$ %, and $37.6 \pm 8.8/12.5 \pm 2.6$ %, correspondingly, while in sham control specimens routine incubation of the sample without adding mast cell activators did not induce any significant changes.

Trends similar to that described above were observed with analysis of the mean pixel intensities of mast cells in conditions before and after treatment of 9- and 24-month-old mesenteric segments by various activators. Mean pixel intensity of stained cells gives a measure of the degree of activation of mast cells as a result of treatment. Since increased number of mesenteric mast cells was found to be activated in basal conditions in 24-month-old segments, the mean pixel intensity of mast cells before treatment is significantly raised compared to 9-month-old segments. After treatment with mast cell activators, the mean pixel intensity of mast cells increased in both age groups but the change in intensity from untreated to treated conditions is much less in 24-month-old segments as compared to 9-month-old segments. Hence, although the mean pixel intensity after treatment may be more in 24-month-old rats (this is due to the fact that there were already more activated cells in basal conditions), the change in mean intensity as a result of treatment is significantly less in old rats.

Aging is considered to be a chronic inflammatory process with a shift towards a proinflammatory cytokine profile in tissues that may account for numerous deleterious vascular changes associated with aging (Bruunsgaard et al. 2001; Bruunsgaard and Pedersen 2003; Csiszar et al. 2003). The increased state of pre-activation as well as increased number of mast cells in aged mesentery, which we at the first time confirmed in this study, indicates existence of chronic inflammatory environment in mesentery since activated mast cells would have released their preformed inflammatory mediators like histamine, proteases, and cytokines like TNF alpha, etc. Such aging-associated chronic inflammatory environment, sign of which is an increased state of activation of aged mast cells in close proximity to aged MLV, may be one of the important causes of alterations in lymphatic pump function in aged animals. Previously we demonstrated that application of L-NAME in situ caused an abrupt and significant increase in contraction frequency of aged MLVs but not in isolated vessels (Akl et al. 2011). This proves that in the immediate vicinity of aged MLV, there are some counterbalancing mediators which have effects opposite to the lymphatic inhibitory effects of NO in aged rats. Such counterbalancing effects do not exist in isolated MLV preparations (Nagai et al. 2011) because carefully dissected lymphatic vessels do not have mast cells in their walls and surrounding tissues are eliminated during vessel isolation procedure. The presence of increased numbers of activated mast cells by aged MLVs indirectly confirms the existence of the increased local concentrations of histamine and other mediators which are known to influence lymphatic contractility and in particular are able to induce positive chronotropic effect on lymphatic vessels (discussed above). These mediators have potential to further increase contraction frequency of aged MLV in time when the inhibitory effect of NO is removed by L-NAME administration (Akl et al. 2011). In light of our present findings, there is in average ~400 % increase in the number of activated mast cells in aged mesentery in addition to 27 % increase in their density near aged lymphatic vessels, we may consider the cumulative action of mast cell mediators chronically released from activated mast cells in resting conditions as an important mechanism of age-induced shifts in lymphatic contractile function. Further detailed investigations are necessary to determine how different mast cell-derived mediators and by which mechanisms influence function of the aged lymphatic vessels.

Although there has been a great deal of work done on inflammation in vascular aging, there is lack of sufficient research which may apply to the aging of the lymphatic system. In spite of a chronic inflammatory state, aging has been thought to involve immunosenescence with deficient innate as well as adaptive immune response (Gomez et al. 2005, 2008; Solana et al. 2006; Nomellini et al. 2008). Mast cells are considered to be the first line of defense against allergens and pathogens and thus are important initiators of innate immune response (Marshall 2004; Dawicki and Marshall 2007; Mekori and Metcalfe 2000; Marshall et al. 2003; Malaviya and Georges 2002). Various inflammatory and chemotactic molecules produced and released from mast cells in response to foreign allergens or pathogens help recruitment of other antigen-presenting cells at the site of inflammation and participate in shaping the host-adaptive response (Raible et al. 1992; Ramos et al. 2003; Heib et al. 2007; Dawicki et al. 2010). Historically studies on mast cells in aging have been limited to dermal and lung mast cells in context of studies on allergy and asthma (Migally et al. 1983; Hart et al. 1999; Gunin et al. 2011). In this study we have, for the first time, analyzed the aging-associated changes in mast cell number and their behavior in rat mesentery near lymphatic vessels. In this study we also have compared the responses of mesenteric mast cells to different acute stimuli in adult and aged animals. We used known mast cell activators: substance P—an inflammatory neurotransmitter; anti-rat DNP IgE—an allergic mediator; peptidoglycan from *S. aureus* (PGN)—a classical toll-like receptor 2 ligand, and compound 48/80—standard chemical activator to stimulate mast cells ex vivo in live mesenteric tissue containing MLV. Unlike in vitro experiments where the total number of cells is arbitrary, our current ex vivo preparation helped us to evaluate the number of pre-activated cells as well as number of intact cells available in the particular tissue bed to be newly activated by acute inflammatory stimuli. We found that in spite of the fact that 24-month-old rats had ~27 % increase in total mast cell number in the mesenteric bed; the number of intact cells available to react to the presence of acute stimuli is significantly decreased in aged rats. We predict that the presence of extensive degranulation of mast cells in basal conditions in 24-month-old rats promotes a chronic inflammatory environment but the reduced availability of intact mast cells to react to acute noxious stimuli is one of the major contributory reasons of delayed immune response to acute inflammation in elderly. We discovered that a smaller fraction of previously nonactivated mast cells were activated as a result of either biological or chemical stimulation in 24-month-old rats as compared to 9-month-old animals. However in case of IgE stimulation, the diminished ability of aged mast cells to be activated (lower number of activated mast cells after treatment) also has its own impact on age-associated alterations in functional status of mast cells and in their acute response to acute stimulation. The mean pixel intensity as a result of treatment (change of intensity after treatment from before treatment conditions) is always significantly less in 24-month-old rats compared to 9-month-old rats. The presence of extensive degranulation in basal conditions in aged mesentery increased the mean pixel intensity of mesenteric mast cells after treatment in 24-month-old rats as compared to 9-month-old rats but the increase in mean pixel intensity as a result of activation due to treatment was always significantly more in young rats as compared to old rats. Taken together our findings indicate that in aging, mesenteric mast cells have reduced the ability to be activated by acute inflammatory stimuli. While deficient immune cell function has been reported in aging (Gomez et al. 2005, 2008; Solana et al. 2006; Nomellini et al. 2008), there is no detailed information on the aging-associated alterations of mast cells and how they affect the function of MLV. The limited number of aged mast cells located in the mesentery to react to the presence of acute stimuli may be considered contributory to the aging-associated deteriorations in immune response. Further investigations will be able to answer the important questions on the mechanisms of such effects discovered in this study (Chatterjee and Gashev 2012).

The free radical theory of aging suggests that there is a progressive reduction in NO production and a simultaneous increase in free radical production leading to endothelial dysfunction. Beckman and Ames suggested that oxidative stress is an important factor contributing to vascular dysfunction with aging (Beckman and Ames 1998). Vascular aging is associated with both structural and functional changes that can take place at the level of the endothelium, smooth muscle cells, and the extracellular matrix of vessels. Aging is also correlated with increased oxidative stress and oxidative damage, and the endothelium appears to be an important source of superoxide anion ($O_2^{\bullet -}$) in the vascular wall. Marin and Rodriguez-Martinez reported (1995) that endothelial cells are vulnerable to oxidative stress due to their low antioxidant capacity. Aging-related endothelial dysfunction may involve mechanisms such as alterations in the antioxidant defense systems, increased oxidative injury, or both. Studies suggest that inactivation of nitric oxide by superoxide contributes to impaired vascular function (Mayhan 1997; Didion et al. 2001, 2002a, b). NO reacts with superoxide radical ($O_2^{\bullet-}$) to form peroxynitrite ($ONOO^-$), which can further induce protein modification and DNA damage in the microvascular system (Pacher et al. 2007). Thus, decrease in NO bioavailability as a result of excess $O_2^{\bullet-}$ formation is a major cause of endothelial dysfunction in aging. Aging-associated elevations in oxidative stress may be related to alterations in antioxidant defense enzymes such as the superoxide dismutase (SOD) isoforms Cu/Zn-SOD (located in the cytoplasm), Mn-SOD (located in the mitochondria), and extracellular-SOD (EC-SOD). Additionally, the oxidative stress may come from increased production of reactive oxygen species (ROS) via mitochondrial dysfunction, activation of NADPH oxidase, or uncoupling of NOS (Chu et al. 2005; Brown et al. 2006, 2007). The balance between the levels of free radical production and cellular antioxidant activity determine the oxidative stress on the tissue and subsequent degree of oxidative damage. Importantly, Zawieja et al. reported (1991) that oxygen radicals significantly inhibited contractile activity of rat MLV, thus we propose that aging-associated increases in oxidative stress could contribute to the decline in contractile activity seen in aged lymphatic vessels. Currently, there are no investigations of aging-associated oxidative stress or oxidative damage in lymphatic vessels. Thus, the goal of the one of our recently published work (Thangaswamy et al. 2012) was to examine the aging-induced changes in expression and activity of the major cellular antioxidant enzyme, SOD (Cu/Zn-SOD, EC-SOD, and Mn-SOD isoforms) while also evaluating peroxynitrite-mediated cellular damage and mitochondria-related superoxide radical production in aged MLV from F-344 rats in comparison with their adult counterparts.

In discussed study (Thangaswamy et al. 2012) we measured total activity of SOD enzyme activity in adult and aged MLV and found a significant decrease in 24-month-old MLV (3.98 ± 0.08 Unit/mL) compared to 9-month-old vessels (4.36 ± 0.04 Unit/mL). Additionally, we analyzed TBARS as an indicator of lipid peroxidation and found that the concentration of TBARS was significantly elevated in 24-month-old MLV (53.7 ± 10.8 nM/mg wet tissue) when compared to 9-month-old vessels (3.4 ± 0.8 nM/mg wet tissue). Next we performed the determination of total cellular superoxide and mitochondrial ROS production in MLV by measuring the total levels of cellular superoxide by DHE fluorescence in live adult and aged MLV. After 30 min of incubation with this superoxide dye, numerous fluorescently labeled cells were identified in MLV, and the density of fluorescent signal was significantly higher (~97 %) in the 24-month-old MLV (24.7 ± 2.9 arbitrary units) when compared to 9-month-old vessels (12.5 ± 0.8 arbitrary units). Additionally, we assessed mitochondrial ROS production using the mitochondria-specific ROS-sensitive fluorescent dye MitoTracker Red CM-H2XRos and identified significantly elevated (~61 %) production of mitochondrial ROS in 24-month-old MLV (21.3 ± 1.9 arbitrary units) compared to 9-month-old vessels (13.2 ± 2.1 arbitrary units).

Furthermore, we performed western blot analyzes to compare the protein expression of SOD isoforms (Cu/Zn-SOD, Mn-SOD, and EC-SOD) and nitro-tyrosine (as an indicator of oxidative damage) in 9- and 24-month-old MLV. We found that the Cu/Zn-SOD isoform was expressed significantly lower (~51 %) in 24-month-old

MLV (normalized signal intensity (NSI): 2.53±0.33) compared to 9-month-old MLV (NSI: 5.16±0.53) while the EC-SOD isoform was expressed significantly higher (~78 %) in 24-month-old MLV (NSI: 5.42±0.53) compared to 9-month-old MLV (NSI: 3.04±0.40). We did not find significant differences in the expression of the Mn-SOD protein isoform between 9- and 24-month-old MLV samples (NSI: 0.96±0.18 vs. 1.60±0.31, respectively). Protein bound nitro-tyrosine formation is commonly used to demonstrate nitric oxide-dependent oxidative damage. In this study, we examined the quantity of nitro-tyrosine formation in 9- and 24-month-old MLV via western blot. We observed a significant increase (~148 %) in the expression of nitro-tyrosine protein corresponding to 215 kDa band in 24-month-old MLV (NSI: 1.05±0.18) when compared to 9-month-old MLV (NSI: 0.42±0.14). Conversely, we did not observe significant differences in the 36 kDa band between age groups.

We also performed immunohistochemical labeling of SOD isoforms (Cu/Zn-SOD, Mn-SOD, and EC-SOD) and nitro-tyrosine (as an indicator of oxidative damage) in 9- and 24-month-old MLV. We found that the signal intensity (relative to background) for Cu/Zn-SOD isoform was significantly lower (~28 %) in 24-month-old MLV (relative signal intensity (RSI): 6.31±1.32) compared to 9-month-old MLV (RSI: 8.73±1.64). At the same time, we did not find significant differences in the signal intensities of Mn-SOD and EC-SOD isoforms between 9- and 24-month-old MLV samples (RSI: 9.93±4.86 vs. 6.87±1.40 and 11.54±1.33 vs. 11.87±2.67, respectively). We also examined the signal intensity of nitro-tyrosine in 9- and 24-month-old MLV via immunohistochemical labeling to demonstrate NO-dependent oxidative protein damage. We observed a significant increase (~47 %) in signal of nitro-tyrosine labeling in 24-month-old MLV (RSI: 10.67±0.44) when compared to 9-month-old vessels (RSI: 7.24±0.43).

Aging is commonly defined as a functional loss over time that is accompanied by an inability to withstand stress or insult. We recently demonstrated that, as with many other organs, there is an aging-associated decline in the contractile capacity of lymphatic vessels (Gasheva et al. 2007; Nagai et al. 2011); however, the cellular events responsible for this functional loss remain undefined. The role of ROS such as $O_2^{\bullet-}$ in the aging process was initially proposed as the "free radical theory" by Harman in the 1950s (Harman 1956), whereby ROS damage the cellular constituents resulting in a functional decline of the organ systems finally leading to cell death. The participation of ROS is well documented in many pathological conditions typical for the elderly such as cardiovascular diseases, diabetes, cancer, and arthritis. Fridovich (1978) reported that under normal conditions, intracellular $O_2^{\bullet-}$ concentrations are kept at low levels because eukaryotic cells contain large amounts of SOD ($4–10\times10^{-6}$ M). There are numerous reports suggesting that ROS at moderate concentrations act as signaling molecules and play an important role in the regulation of various vascular cellular functions (Cai et al. 2002; Waypa et al. 2002; Miura et al. 2003; Felty et al. 2005; Kimura et al. 2005). Acutely, ROS overproduction can interfere with important signaling cascades such as inactivation of bioavailable NO, alterations in prostaglandin metabolism, or deregulation of calcium and phosphorylation cascades (Vanhoutte 2009a, b). Chronically, ROS overproduction can lead to irreversible oxidation and accumulation of oxidized biological macromolecules, e.g., increase in DNA mutations (Barja and Herrero 2000). Despite the crucial involvement of lymph transport vessels in numerous physiological and pathological processes, there are no reports available on the involvement of oxidative stress in the aging of lymphatic vessels. In this present study, we provided the first evidence for aging-associated elevations in MLV $O_2^{\bullet-}$ and oxidative cellular damage.

Endothelial cell membrane damage is thought to be an early event leading to microvascular dysfunction and may be initiated by several factors including lipid peroxidation. ROS such as hydrogen peroxide, superoxide, and hydroxyl radicals damage biomembranes and induce peroxidation of lipids leading to an increase in cell permeability and loss of endothelial integrity. In this current study, we observed a significant increase in the concentration of TBARS in MLV of 24-month-old compared to 9-month-old rats. We considered this observed increase in the levels of TBARS to be an indicator of elevated aging-associated free radical-induced lipoperoxidation in MLV that could ultimately be associated with cellular membrane damage. These findings correlate with observations by Ohkuma (1989) who proposed that interstitial accumulation of lipoperoxide (a breakdown product of cell membranes) may be potent toxic factor responsible for trophic changes associated with chronic lymphedema (a pathological condition also associated reduced lymphatic contractility (Olszewski 2002; Modi et al. 2007; Olszewski 2008)). In 1993 Ohkuma suggested (Ohkuma 1993) that during the impaired lymph drainage, lipoperoxides, which are normally transported in lymph, may be deposited in the skin and contribute to the soft tissue changes characteristic of chronic lymphedema. In respect to Dr. Ohkuma findings and given the role of mesenteric lymphatics in transporting dietary lipids, we propose that aging-associated lipoperoxidative damage in the mesenteric lymphatic network observed in the present study and the impairment of the mesenteric lymph transport function in aged vessels reported by us earlier (Nagai et al. 2011; Akl et al. 2011) may predispose the elderly to excessive mesenteric fat deposition thereby potentially contributing to the lipid dysregulation commonly seen in aging-associated diseases. Potential inflammatory-related breakdown of antioxidant systems in MLV might be considered as an important trigger in the development of metabolic syndrome even at earlier stages of the life span, which requires further detailed investigations.

As discussed above, diminished NO bioavailability due to increased $O_2^{\bullet-}$ production is one of the major mechanisms responsible for the impaired endothelium-dependent vasodilator responses observed during aging. The interaction between NO and superoxide depletes NO bioactivity thereby altering several key vascular functions of which NO is a pivotal mediator including regulation of smooth muscle tone, platelet activation, and vascular cell signaling (Cai et al. 2002; Vanhoutte 2009; Vanhoutte et al. 2009). In currently discussed study (Thangaswamy et al. 2012), we report the first evidence of a significant increase in superoxide production in live aged MLV thus confirming the existence of the aging-associated oxidative stress in MLV, which may be a key factor in the cascade of events occurring in the mesenteric lymphatic network as a biological consequence of aging.

In normal physiological conditions, antioxidant enzymes prevent the detrimental effects of $O_2^{\bullet -}$. In studying the role of antioxidant defenses in aging, attention has been given to the role of SOD, which efficiently and specifically catalyzes the dismutation of $O_2^{\bullet -}$ to H_2O_2 and O_2. SODs are also involved in the modulation of NO bioactivity. Normally, tissues express three isoforms of SODs including Cu/Zn-SOD (SOD1), Mn-SOD (SOD2), and EC-SOD (EC-SOD or SOD3). Cu/Zn-SOD is an abundant copper- and zinc-containing cellular protein that is present in the cytosol, nucleus, peroxisomes, and mitochondrial inner membrane. Its primary function is to lower the intracellular steady-state concentration of $O_2^{\bullet -}$ (Liochev and Fridovich 2010) Mn-SOD is a mitochondrial enzyme that disposes of $O_2^{\bullet -}$ generated by respiratory chain activity. It can be induced to protect against pro-oxidant insults. Conversely, Mn-SOD activity is decreased in physiologic aging and in diseases such as cancer, asthma, and transplant rejection (Macmillan-Crow and Cruthirds 2001). EC-SOD plays an important role in regulating blood pressure and vascular contraction at least in part through modulating the endothelial function by controlling the levels of extracellular $O_2^{\bullet -}$ and nitric oxide bioactivity in the vasculature (Gongora et al. 2006; Jung et al. 2003). Aging has been shown to produce alterations in the expression and activity of SOD in several tissues (Hollander et al. 2000; Santa Maria et al. 1996). In blood vessels, SOD activity and/or expression has been shown to be altered with aging (Csiszar et al. 2002; Sun et al. 2004). Zawieja et al. (1991) reported that in young MLV, decreases in ejection fraction, contraction frequency, and lymph pump flow caused by high dose superoxide anion treatment were attenuated by SOD application. In our study (Thangaswamy et al. 2012), we observed that total SOD activity was significantly decreased in the aged MLV compared to adult vessels. This result may be a contributing factor to the elevated ROS in aged vessels and/or may augment the oxidative stress on the vessels during aging leading to elevated levels of oxidative damage. In parallel with data obtained by Zawieja et al. (1991), we propose that diminished SOD enzyme activity may play an important role in the development of the aging-associated impairment of the mesenteric lymph transport function reported by us earlier (Nagai et al. 2011; Akl et al. 2011). However, further functional investigations are necessary to confirm the role of SOD protein dysfunction during aging in MLV.

The relative expression of different SOD isoforms in cells and tissues has been investigated extensively and provides clues as to the sources of $O_2^{\bullet -}$ in pathophysiologic states. Cu/Zn-SOD is the predominant isoform in microvessels where it may scavenge $O_2^{\bullet -}$ to increase the bioavailability of NO, which in turn improves endothelium-dependent vascular function. Didion et al. (2002b) suggested that the release of NO from the endothelium is dependent on Cu/Zn-SOD, whereas EC-SOD activity is thought to be required for the protection of NO as it diffuses through the vascular wall. In both large arteries and microvessels, deficiency in Cu/Zn-SOD results in increased levels of vascular superoxide and peroxynitrite, increased myogenic tone, augmented vasoconstrictor responses, and impaired endothelium-dependent NO-mediated relaxation (Didion et al. 2002a; Cooke and Davidge 2003; Veerareddy et al. 2004). In the current study, we confirmed by western blot analyses and immunohistochemical labeling significantly lower levels of Cu/Zn-SOD isoform expression in 24-month-old compared to 9-month-old MLV, such changes may be a contributor to the aging-associated impaired endothelium-dependent NO-mediated regulation on the lymphatic vessels (Gasheva et al. 2007; Nagai et al. 2011; Akl et al. 2011). While we did observe an increase EC-SOD isoform expression in 24-month-old compared to 9-month-old MLV via western blot, we were not able to see any difference in expression of EC-SOD via immunohistochemical labeling. While these differing results for EC-SOD may simply reflect the detection abilities of these two techniques, it is important to note that total SOD activity in 24-month-old MLV is still decreased relative to the 9-month-old vessels. Thus, even if EC-SOD protein expression is indeed elevated in the 24-month-old vessels (as indicated via western blot), the total antioxidant activity of the aged vessels is still depressed relative to the adult vessels presumably due to the consistent depression in Cu/Zn-SOD protein expression. Follow-up studies are needed to assess the specific importance of the Cu/Zn-SOD isoform depletion and the possible EC-SOD enrichment in the development of the aging-associated oxidative damage in lymphatic vessels.

The increased levels of $O_2^{\bullet-}$ in aged MLV may result in increased NO scavenging and subsequent $ONOO^-$ formation, which is known to initiate oxidative modification of proteins ultimately leading to lipid peroxidation or DNA damage (Beckman and Ames 1998; Beckman et al. 1992; Wei et al. 2001). It is known that peroxynitrite at submicromolar concentrations causes the nitration of protein-bound tyrosine residues in Mn-SOD or prostacyclin synthase (Goldstein et al. 2000). Our data clearly demonstrate significantly increased nitro-tyrosine levels in aged MLV, thus indicating increased formation of $ONOO^-$. This last observation suggests that there may be diminished levels of basal NO in the vessel due to the inability of SOD to completely scavenge the elevated $O_2^{\bullet-}$ formation during aging. Consequently, the loss of NO bioactivity associated with increased vascular $O_2^{\bullet-}$ in the MLV during aging may play a potentially important role in the pathogenesis of lymphatic endothelial dysfunction. We also found that mitochondrial ROS increased in the aged MLV compared to their adult counterparts. This also might enhance the reaction between $O_2^{\bullet-}$ and NO. It is demonstrated (Aulak et al. 2004) that protein tyr-nitration mediated by $ONOO^-$ increases with age and ultimately leads to the inhibition of mitochondrial energy production by causing site-specific lesions in the electron transport chain, in enzymes involved in the citric acid cycle, or in enzymes necessary for energy transfer. In connection with these findings, aging-associated weakening of lymphatic pumping and disturbances in its NO-dependent regulatory mechanisms (Gasheva et al. 2007; Nagai et al. 2011; Akl et al. 2011) may be linked to increased nitro-tyrosine levels in aged lymphatic vessels.

In conclusion, in this comprehensive review of recent research findings related to the nature and the mechanisms of aging-associated alterations of lymphatic contractility and flow, we provided extended summary of current knowledge in this field of lymphatic biology. Although first critical steps on the way of discovery of the aging-associated alterations of lymphatic contractility and pumping already performed, these new findings opened new important horizons for new scientific endeavors. We hope that future continuation of the research efforts in this area will provide not only novel fundamental knowledge on the biology of lymphatic aging, but also will create solid foundation for the subsequent developments of the lymphatic-oriented therapeutic interventions during many diseases in elderly.

References

Akl TJ et al (2011) Mesenteric lymph flow in adult and aged rats. Am J Physiol Heart Circ Physiol 301(5):H1828–H1840

Allen JM, Iggulden HL, McHale NG (1986) Beta-adrenergic inhibition of bovine mesenteric lymphatics. J Physiol 374:401–411

Amin K (2012) The role of mast cells in allergic inflammation. Respir Med 106(1):9–14

Armenio S et al (1981a) Spontaneous contractility in the human lymph vessels. Lymphology 14(4):173–178

Armenio S et al (1981b) Spontaneous contractility of the lymphatic vessels in man. Angilogia 33(6):325–327 [Spanish]

Aulak KS et al (2004) Dynamics of protein nitration in cells and mitochondria. Am J Physiol Heart Circ Physiol 286(1):H30–H38

Barja G, Herrero A (2000) Oxidative damage to mitochondrial DNA is inversely related to maximum life span in the heart and brain of mammals. FASEB J 14(2):312–318

Beckman KB, Ames BN (1998) The free radical theory of aging matures. Physiol Rev 78(2):547–581

Beckman JS et al (1992) Kinetics of superoxide dismutase- and iron-catalyzed nitration of phenolics by peroxynitrite. Arch Biochem Biophys 298(2):438–445

Beil WJ, Schulz M, Wefelmeyer U (2000) Mast cell granule composition and tissue location—a close correlation. Histol Histopathol 15(3):937–946

Benoit JN (1991) Relationships between lymphatic pump flow and total lymph flow in the small intestine. Am J Physiol 261(6 Pt 2):H1970–H1978

Benoit JN, Zawieja DC (1994) Gastrointestinal lymphatics. In: Johnson L (ed) Physiology of the gastrointestinal tract. Raven, New York, pp 1669–1692

Benoit JN et al (1989) Characterization of intact mesenteric lymphatic pump and its responsiveness to acute edemagenic stress. Am J Physiol 257(6 Pt 2):H2059–H2069

Boesiger J et al (1998) Mast cells can secrete vascular permeability factor/ vascular endothelial cell growth factor and exhibit enhanced release after immunoglobulin E-dependent upregulation of fc epsilon receptor I expression. J Exp Med 188(6):1135–1145

Bohlen HG et al (2009) Phasic contractions of rat mesenteric lymphatics increase basal and phasic nitric oxide generation in vivo. Am J Physiol Heart Circ Physiol 297(4):H1319–H1328

Bridenbaugh EA, Gashev AA, Zawieja DC (2003) Lymphatic muscle: a review of contractile function. Lymphat Res Biol 1(2):147–158

Brill GE et al (2001) Functional organization of lymphatic microvessels of the rat mesentery. Ross Fiziol Zh Im I M Sechenova 87(5):600–607 [In Russian]

Brown KA et al (2006) Gene transfer of extracellular superoxide dismutase protects against vascular dysfunction with aging. Am J Physiol Heart Circ Physiol 290(6):H2600–H2605

Brown KA et al (2007) Effect of aging, MnSOD deficiency, and genetic background on endothelial function: evidence for MnSOD haploinsufficiency. Arterioscler Thromb Vasc Biol 27(9):1941–1946

Bruunsgaard H, Pedersen BK (2003) Age-related inflammatory cytokines and disease. Immunol Allergy Clin North Am 23(1):15–39
Bruunsgaard H, Pedersen M, Pedersen BK (2001) Aging and proinflammatory cytokines. Curr Opin Hematol 8(3):131–136
Bulekbaeva LE (1988) The volume rate of lymph flow in dogs in postnatal ontogeny. Zh Evol Biokhim Fiziol 24(4):599–600 [In Russian]
Cai H et al (2002) NAD(P)H oxidase-derived hydrogen peroxide mediates endothelial nitric oxide production in response to angiotensin II. J Biol Chem 277(50):48311–48317
Chatterjee V, Gashev AA (2012) Aging-associated shifts in functional status of mast cells located by adult and aged mesenteric lymphatic vessels. Am J Physiol Heart Circ Physiol 303(6):H693–H702
Chevalier S, Ferland G, Tuchweber B (1996) Lymphatic absorption of retinol in young, mature, and old rats: influence of dietary restriction. FASEB J 10(9):1085–1090
Chu Y et al (2005) Vascular effects of the human extracellular superoxide dismutase R213G variant. Circulation 112(7):1047–1053
Cooke CL, Davidge ST (2003) Endothelial-dependent vasodilation is reduced in mesenteric arteries from superoxide dismutase knockout mice. Cardiovasc Res 60(3):635–642
Crowe MJ et al (1997) Co-ordination of contractile activity in guinea-pig mesenteric lymphatics. J Physiol 500(Pt 1):235–244
Csiszar A et al (2002) Aging-induced phenotypic changes and oxidative stress impair coronary arteriolar function. Circ Res 90(11):1159–1166
Csiszar A et al (2003) Aging-induced proinflammatory shift in cytokine expression profile in coronary arteries. FASEB J 17(9):1183–1185
Davis MJ et al (2009) Myogenic constriction and dilation of isolated lymphatic vessels. Am J Physiol Heart Circ Physiol 296(2):H293–H302
Dawicki W, Marshall JS (2007) New and emerging roles for mast cells in host defence. Curr Opin Immunol 19(1):31–38
Dawicki W et al (2010) Mast cells, histamine, and IL-6 regulate the selective influx of dendritic cell subsets into an inflamed lymph node. J Immunol 184(4):2116–2123
Detoraki A et al (2009) Vascular endothelial growth factors synthesized by human lung mast cells exert angiogenic effects. J Allergy Clin Immunol 123(5):1142-9–1149.e1-5
Didion SP, Hathaway CA, Faraci FM (2001) Superoxide levels and function of cerebral blood vessels after inhibition of CuZn-SOD. Am J Physiol Heart Circ Physiol 281(4):H1697–H1703
Didion SP et al (2002a) Increased superoxide and vascular dysfunction in CuZnSOD-deficient mice. Circ Res 91(10):938–944
Didion SP et al (2002b) Superoxide contributes to vascular dysfunction in mice that express human renin and angiotensinogen. Am J Physiol Heart Circ Physiol 283(4):H1569–H1576
Dixon JB et al (2006) Lymph flow, shear stress, and lymphocyte velocity in rat mesenteric prenodal lymphatics. Microcirculation 13(7):597–610
Dobbins DE (1998) Receptor mechanisms of serotonin-induced prenodal lymphatic constriction in the canine forelimb. Am J Physiol 274(2 Pt 2):H650–H654
Dobbins DE, Buehn MJ, Dabney JM (1990) Constriction of perfused lymphatics by acetylcholine, bradykinin and histamine. Microcirc Endothelium Lymphatics 6(6):409–425
Esposito P et al (2002) Corticotropin-releasing hormone and brain mast cells regulate blood–brain-barrier permeability induced by acute stress. J Pharmacol Exp Ther 303(3):1061–1066
Fedosov IV et al (2002) Recording of lymph flow dynamics in microvessels using correlation properties of scattered coherent radiation. Quantum Electronics 32(11):970–974
Felty Q et al (2005) Estrogen-induced mitochondrial reactive oxygen species as signal-transducing messengers. Biochemistry 44(18):6900–6909
Ferguson MK, Shahinian HK, Michelassi F (1988) Lymphatic smooth muscle responses to leukotrienes, histamine and platelet activating factor. J Surg Res 44(2):172–177
Fox JLR, von der Weid P-Y (2002) Effects of histamine on the contractile and electrical activity in isolated lymphatic vessels of the guinea-pig mesentery. Br J Pharmacol 136(8):1210–1218

Fridovich I (1978) Superoxide dismutases: defence against endogenous superoxide radical. Ciba Found Symp 65:77–93
Galanzha EI et al (2002) Nitric oxide in the lymphatic microvessel regulation. Ross Fiziol Zh Im I M Sechenova 88(8):983–989 [In Russian]
Galanzha EI, Tuchin VV, Zharov VP (2005) In vivo integrated flow image cytometry and lymph/blood vessels dynamic microscopy. J Biomed Opt 10(5):054018
Galli SJ, Nakae S, Tsai M (2005) Mast cells in the development of adaptive immune responses. Nat Immunol 6(2):135–142
Gashev AA (1989) The pump function of the lymphangion and the effect on it of different hydrostatic conditions. Fiziol Zh SSSR Im I M Sechenova 75(12):1737–1743 [In Russian]
Gashev AA (1991) The mechanism of the formation of a reverse fluid filling in the lymphangions. Fiziol Zh SSSR Im I M Sechenova 77(7):63–69 [In Russian]
Gashev AA (2002) Physiologic aspects of lymphatic contractile function: current perspectives. Ann N Y Acad Sci 979:178–187, discussion 188–96
Gashev AA (2008) Lymphatic vessels: pressure- and flow-dependent regulatory reactions. Ann N Y Acad Sci 1131:100–109
Gashev AA (2010) Basic mechanisms controlling lymph transport in the mesenteric lymphatic net. Ann N Y Acad Sci 1207(Suppl 1):E16–E20
Gashev AA, Zawieja DC (2001) Physiology of human lymphatic contractility: a historical perspective. Lymphology 34(3):124–134
Gashev AA, Zawieja DC (2010) Hydrodynamic regulation of lymphatic transport and the impact of aging. Pathophysiology 17(4):277–287
Gashev AA et al (1990) The mechanisms of lymphangion interaction in the process of the lymph movement. Fiziol Zh SSSR Im I M Sechenova 76(11):1489–1508 [In Russian]
Gashev AA, Davis MJ, Zawieja DC (2002) Inhibition of the active lymph pump by flow in rat mesenteric lymphatics and thoracic duct. J Physiol 540(Pt 3):1023–1037
Gashev AA et al (2004) Regional variations of contractile activity in isolated rat lymphatics. Microcirculation 11(6):477–492
Gashev AA, Delp MD, Zawieja DC (2006) Inhibition of active lymph pump by simulated microgravity in rats. Am J Physiol Heart Circ Physiol 290(6):H2295–H2308
Gasheva OY, Zawieja DC, Gashev AA (2006) Contraction-initiated NO-dependent lymphatic relaxation: a self-regulatory mechanism in rat thoracic duct. J Physiol 575(Pt 3):821–832
Gasheva OY et al (2007) Age-related alterations of active pumping mechanisms in Rat thoracic duct. Microcirculation 14(8):827–839
Goldstein S et al (2000) Tyrosine nitration by simultaneous generation of (.)NO and O-(2) under physiological conditions. How the radicals do the job. J Biol Chem 275(5):3031–3036
Gomez CR, Boehmer ED, Kovacs EJ (2005) The aging innate immune system. Curr Opin Immunol 17(5):457–462
Gomez CR et al (2008) Innate immunity and aging. Exp Gerontol 43(8):718–728
Gongora MC et al (2006) Role of extracellular superoxide dismutase in hypertension. Hypertension 48(3):473–481
Grutzkau A et al (1998) Synthesis, storage, and release of vascular endothelial growth factor/vascular permeability factor (VEGF/VPF) by human mast cells: implications for the biological significance of VEGF206. Mol Biol Cell 9(4):875–884
Gunin AG et al (2011) Age-related changes in proliferation, the numbers of mast cells, eosinophils, and cd45-positive cells in human dermis. J Gerontol A Biol Sci Med Sci 66(4):385–392
Hargens AR, Zweifach BW (1977) Contractile stimuli in collecting lymph vessels. Am J Physiol 233(1):H57–H65
Harman D (1956) Aging: a theory based on free radical and radiation chemistry. J Gerontol 11(3):298–300
Hart PH et al (1999) Age-related changes in dermal mast cell prevalence in BALB/c mice: functional importance and correlation with dermal mast cell expression of kit. Immunology 98(3):352–356

Harvima IT, Nilsson G (2011) Mast cells as regulators of skin inflammation and immunity. Acta Derm Venereol 91(6):644–650

Hayes H et al (2003) Development and characterization of endothelial cells from rat microlymphatics. Lymphat Res Biol 1(2):101–119

Heib V et al (2007) Mast cells are crucial for early inflammation, migration of Langerhans cells, and CTL responses following topical application of TLR7 ligand in mice. Blood 110(3):946–953

Hollander D, Dadufalza V (1990) Influence of aging on vitamin A transport into the lymphatic circulation. Exp Gerontol 25(1):61–65

Hollander J et al (2000) Superoxide dismutase gene expression in skeletal muscle: fiber-specific effect of age. Mech Ageing Dev 116(1):33–45

Johnston MG, Kanalec A, Gordon JL (1983) Effects of arachidonic acid and its cyclo-oxygenase and lipoxygenase products on lymphatic vessel contractility in vitro. Prostaglandins 25(1):85–98

Jung O et al (2003) Extracellular superoxide dismutase is a major determinant of nitric oxide bioavailability: in vivo and ex vivo evidence from ecSOD-deficient mice. Circ Res 93(7):622–629

Kimura S et al (2005) Role of NAD(P)H oxidase- and mitochondria-derived reactive oxygen species in cardioprotection of ischemic reperfusion injury by angiotensin II. Hypertension 45(5):860–866

Kluth W (1951) The presence of mast cells in the thoracic duct. Zentralbl Allg Pathol 87(4–5):139–141

Kwon S, Sevick-Muraca EM (2007) Noninvasive quantitative imaging of lymph function in mice. Lymphat Res Biol 5(4):219–231

Liochev SI, Fridovich I (2010) Mechanism of the peroxidase activity of Cu, Zn superoxide dismutase. Free Radic Biol Med 48(12):1565–1569

Lobov GI, Pan'kova MN (2010) Heparin inhibits contraction of smooth muscle cells in lymphatic vessels. Bull Exp Biol Med 149(1):4–6

Lorenz D et al (1998) Mechanism of peptide-induced mast cell degranulation. Translocation and patch-clamp studies. J Gen Physiol 112(5):577–591

Lundequist A, Pejler G (2011) Biological implications of preformed mast cell mediators. Cell Mol Life Sci 68(6):965–975

Lytinas M et al (2003) Acute stress results in skin corticotropin-releasing hormone secretion, mast cell activation and vascular permeability, an effect mimicked by intradermal corticotropin-releasing hormone and inhibited by histamine-1 receptor antagonists. Int Arch Allergy Immunol 130(3):224–231

Macmillan-Crow LA, Cruthirds DL (2001) Invited review: manganese superoxide dismutase in disease. Free Radic Res 34(4):325–336

Malaviya R, Abraham SN (2001) Mast cell modulation of immune responses to bacteria. Immunol Rev 179:16–24

Malaviya R, Georges A (2002) Regulation of mast cell-mediated innate immunity during early response to bacterial infection. Clin Rev Allergy Immunol 22(2):189–204

Marchetti C et al (1997) Endothelin and nitric oxide synthase in lymphatic endothelial cells: immunolocalization in vivo and in vitro. Anat Rec 248(4):490–497

Marin J, Rodriguez-Martinez MA (1995) Nitric oxide, oxygen-derived free radicals and vascular endothelium. J Auton Pharmacol 15(4):279–307

Marshall JS (2004) Mast-cell responses to pathogens. Nat Rev Immunol 4(10):787–799

Marshall JS, Jawdat DM (2004) Mast cells in innate immunity. J Allergy Clin Immunol 114(1):21–27

Marshall JS, King CA, McCurdy JD (2003) Mast cell cytokine and chemokine responses to bacterial and viral infection. Curr Pharm Des 9(1):11–24

Mayhan WG (1997) Superoxide dismutase partially restores impaired dilatation of the basilar artery during diabetes mellitus. Brain Res 760(1–2):204–209

McHale NG (1990) Lymphatic innervation. Blood Vessels 27(2–5):127–136
McHale NG (1992) The lymphatic circulation. Ir J Med Sci 161(8):483–486
McHale NG, Meharg MK (1992) Co-ordination of pumping in isolated bovine lymphatic vessels. J Physiol 450:503–512
McHale NG, Roddie IC (1976) The effect of transmural pressure on pumping activity in isolated bovine lymphatic vessels. J Physiol 261(2):255–269
Mekori YA, Metcalfe DD (2000) Mast cells in innate immunity. Immunol Rev 173:131–140
Migally NB et al (1983) Density and ultrastructure of mast cells in lung vessels of aging rats exposed to and recovering from chronic hypoxia. Cell Tissue Res 232(3):601–608
Mislin H (1961) Experimental detection of autochthonous automatism of lymph vessels. Experientia 17:29–30 [In German]
Mislin H (1971) The contractile properties of lymphatic vessels. Angiologica 8(3–5):207–211
Mislin H, Schipp R (1966) Structural and functional relations of the lymph vessels. Progress in lymphology. In: Proceedings of the International symposium on lymphology, Zurich, July 19–23, pp. 360–365.
Miura H et al (2003) Role for hydrogen peroxide in flow-induced dilation of human coronary arterioles. Circ Res 92(2):e31–e40
Mizuno R, Koller A, Kaley G (1998) Regulation of the vasomotor activity of lymph microvessels by nitric oxide and prostaglandins. Am J Physiol 274(3 Pt 2):R790–R796
Modi S et al (2007) Human lymphatic pumping measured in healthy and lymphoedematous arms by lymphatic congestion lymphoscintigraphy. J Physiol 583(Pt 1):271–285
Mousli M et al (1989) Activation of rat peritoneal mast cells by substance P and mastoparan. J Pharmacol Exp Ther 250(1):329–335
Muthuchamy M et al (2003) Molecular and functional analyses of the contractile apparatus in lymphatic muscle. FASEB J 17(8):920–922
Nagai T, Bridenbaugh EA, Gashev AA (2011) Aging-associated alterations in contractility of rat mesenteric lymphatic vessels. Microcirculation 18(6):463–473. doi:10.1111/j.1549-8719.2011.00107.x
Nomellini V, Gomez CR, Kovacs EJ (2008) Aging and impairment of innate immunity. Contrib Microbiol 15:188–205
Ohhashi T, Takahashi N (1991) Acetylcholine-induced release of endothelium-derived relaxing factor from lymphatic endothelial cells. Am J Physiol 260(4 Pt 2):H1172–H1178
Ohhashi T, Yokoyama S (1994) Nitric oxide and the lymphatic system. Jpn J Physiol 44(4):327–342
Ohhashi T, Kawai Y, Azuma T (1978) The response of lymphatic smooth muscles to vasoactive substances. Pflugers Arch 375:183–188
Ohhashi T, Azuma T, Sakaguchi M (1980) Active and passive mechanical characteristics of bovine mesenteric lymphatics. Am J Physiol 239(1):H88–H95
Ohhashi T et al (2005) Current topics of physiology and pharmacology in the lymphatic system. Pharmacol Ther 105(2):165–188
Ohkuma M (1989) Lipoperoxide in dog thoracic duct lymph. Lymphology 22(3):150–152
Ohkuma M (1993) Lipoperoxide in the dermis of patients with lymph stasis. Lymphology 26(1):38–41
Olszewski WL (2002) Contractility patterns of normal and pathologically changed human lymphatics. Ann N Y Acad Sci 979:52–63, discussion 76–9
Olszewski WL (2008) Contractility patterns of human leg lymphatics in various stages of obstructive lymphedema. Ann N Y Acad Sci 1131:110–118
Orlov RS, Lobov GI (1984) Ionic mechanisms of the electrical activity of the smooth-muscle cells of the lymphatic vessels. Fiziol Zh SSSR Im I M Sechenova 70(5):712–721 [In Russian]
Orlov RS, Borisov AV, Borisova RP (1983) Lymphatic vessels. Structure and mechanisms of contractile activity. Nauka, Leningrad, p 253
Pacher P, Beckman JS, Liaudet L (2007) Nitric oxide and peroxynitrite in health and disease. Physiol Rev 87(1):315–424

Pan'kova MN et al (2011) [Effects of histamine on contractile activity of lymphatic node capsules. The NO role]. Ross Fiziol Zh Im I M Sechenova 97(6):633–640
Petunov SG et al (2010) Effect of histamine on spontaneous contractions of mesenteric lymphatic vessels and lymph nodes of white rats: endothelium-dependent responses. Dokl Biol Sci 432:176–180 [Translated from Russian]
Plaku KJ, von der Weid PY (2006) Mast cell degranulation alters lymphatic contractile activity through action of histamine. Microcirculation 13(3):219–227
Rabinovitz AJ, Saphir O (1965) The thoracic duct; significance of age-related changes and of lipid in the wall. Circulation 31:899–905
Raible DG et al (1992) Mast cell mediators prostaglandin-D2 and histamine activate human eosinophils. J Immunol 148(11):3536–3542
Ramos CD et al (2003) Neutrophil migration induced by IL-8-activated mast cells is mediated by CINC-1. Cytokine 21(5):214–223
Saban R et al (2002) Mast cells mediate substance P-induced bladder inflammation through an NK(1) receptor-independent mechanism. Am J Physiol Renal Physiol 283(4):F616–F629
Salamon P et al (2005) Human mast cells release Interleukin-8 and induce neutrophil chemotaxis on contact with activated T cells. Allergy 60(10):1316–1319
Santa Maria C, Ayala A, Revilla E (1996) Changes in superoxide dismutase activity in liver and lung of old rats. Free Radic Res 25(5):401–405
Schmid-Schonbein GW (1990) Microlymphatics and lymph flow. Physiol Rev 70(4):987–1028
Schwartz LB (1987) Mediators of human mast cells and human mast cell subsets. Ann Allergy 58(4):226–235
Shakoory B et al (2004) The role of human mast cell-derived cytokines in eosinophil biology. J Interferon Cytokine Res 24(5):271–281
Shirasawa Y, Ikomi F, Ohhashi T (2000) Physiological roles of endogenous nitric oxide in lymphatic pump activity of rat mesentery in vivo. Am J Physiol 278(4):G551–G556
Simon T, Laszlo V, Falus A (2011) Impact of histamine on dendritic cell functions. Cell Biol Int 35(10):997–1000
Sinzinger H, Kaliman J, Mannheimer E (1984) Regulation of human lymph contractility by prostaglandins and thromboxane. Lymphology 17(2):43–45
Solana R, Pawelec G, Tarazona R (2006) Aging and innate immunity. Immunity 24(5):491–494
Sun D et al (2004) Reduced release of nitric oxide to shear stress in mesenteric arteries of aged rats. Am J Physiol Heart Circ Physiol 286(6):H2249–H2256
Suto H et al (2006) Mast cell-associated TNF promotes dendritic cell migration. J Immunol 176(7):4102–4112
Suzuki H et al (1995) Substance P induces degranulation of mast cells and leukocyte adhesion to venular endothelium. Peptides 16(8):1447–1452
Thangaswamy S, Bridenbaugh EA, Gashev AA (2012) Evidence of increased oxidative stress in aged mesenteric lymphatic vessels. Lymphat Res Biol 10(2):53–62
Theoharides TC et al (1998) Corticotropin-releasing hormone induces skin mast cell degranulation and increased vascular permeability, a possible explanation for its proinflammatory effects. Endocrinology 139(1):403–413
Theoharides TC et al (2007) Differential release of mast cell mediators and the pathogenesis of inflammation. Immunol Rev 217:65–78
Unthank J, Bohlen H (1988) Lymphatic pathways and role of valves in lymph propulsion from small intestine. Am J Physiol 254(3 Pt 1):G389–G398
Unthank JL, Hogan RD (1987) The effect of vasoactive agents on the contractions of the initial lymphatics of the bat's wing. Blood Vessels 24:31–44
Van Helden DF (1993) Pacemaker potentials in lymphatic smooth muscle of the guinea-pig mesentery. J Physiol 471:465–479
Vanhoutte PM (2009) Endothelial dysfunction: the first step toward coronary arteriosclerosis. Circ J 73(4):595–601
Vanhoutte PM et al (2009) Endothelial dysfunction and vascular disease. Acta Physiol (Oxf) 196(2):193–222

Veerareddy S et al (2004) Gender differences in myogenic tone in superoxide dismutase knockout mouse: animal model of oxidative stress. Am J Physiol Heart Circ Physiol 287(1):H40–H45

von der Weid P (2001) Review article: lymphatic vessel pumping and inflammation—the role of spontaneous constrictions and underlying electrical pacemaker potentials. Aliment Pharmacol Ther 15(8):1115–1129

von der Weid PY, Zhao J, Van Helden DF (2001) Nitric oxide decreases pacemaker activity in lymphatic vessels of guinea pig mesentery. Am J Physiol 280(6):H2707–H2716

Watanabe N, Kawai Y, Ohhashi T (1988) Dual effects of histamine on spontaneous activity in isolated bovine mesenteric lymphatics. Microvasc Res 36(3):239–249

Waypa GB et al (2002) Mitochondrial reactive oxygen species trigger calcium increases during hypoxia in pulmonary arterial myocytes. Circ Res 91(8):719–726

Wei YH et al (2001) Mitochondrial theory of aging matures—roles of mtDNA mutation and oxidative stress in human aging. Zhonghua Yi Xue Za Zhi (Taipei) 64(5):259–270

Wilhelm DL, Yong LC, Watkins SG (1978) The mast cell: distribution and maturation in the rat. Agents Actions 8(1–2):146–152

Yokoyama S, Ohhashi T (1993) Effects of acetylcholine on spontaneous contractions in isolated bovine mesenteric lymphatics. Am J Physiol 264(5 Pt 2):H1460–H1464

Yong LC, Watkins S, Wilhelm DL (1975) The mast cell: distribution and maturation in the peritoneal cavity of the adult rat. Pathology 7(4):307–318

Yong LC, Watkins SG, Wilhelm DL (1977) The mast cell: II. Distribution and maturation in the peritoneal cavity of the young rat. Pathology 9(3):221–232

Zawieja DC, Davis KL (1993) Inhibition of the active lymph pump in rat mesenteric lymphatics by hydrogen peroxide. Lymphology 26(3):135–142

Zawieja DC et al (1991) Reactive oxygen metabolites inhibit spontaneous lymphatic contractions. Am J Physiol 260(6 Pt 2):H1935–H1943

Zawieja DC et al (1993) Distribution, propagation, and coordination of contractile activity in lymphatics. Am J Physiol 264(4 Pt 2):H1283–H1291

Zawieja DC, Kossman E, Pullin J (1999) Dynamics of the microlymphatic system. J Prog Appl Microcirc 23:100–109

Zweifach B, Prather J (1975) Micromanipulation of pressure in terminal lymphatics in the mesentery. Am J Physiol 228(5):1326–1335

Zweifach B (1972) Micropressure measurements in the terminal lymphatics. In: Ditzel J, Lewis DH (eds) 7th European conference on microcirculation, Aberdeen, Part II Bibl Anat no. 12, pp. 361–365

Chapter 7
Lymph and Lymphatic Capillaries in Cancer

Jacqueline Shields

Abstract Lymph forms from the fluid that is forced out of capillaries and postcapillary venules by hydrostatic pressures, into the interstitium around the vessel. This protein-rich fluid flows through the extracellular matrix and between cells bathing them in nutrients and oxygen and carrying away cellular metabolites and waste products where it is collected by lymphatic capillaries and on to lymph nodes. As with the physiological situation, interstitial fluid and lymph also form within and around tumors, which are collected from cancer-associated tissues. What does change in this situation, however, is the surroundings in which lymph is generated and the tissues exposed to the resulting fluid. The environment in which lymph is formed and transported via can modify its composition and have drastic effects on cells and tissues downstream. This chapter explores the roles of lymphatic function, lymph transport, and their far-reaching implications in cancer development and progression. We pay particular attention to the mechanisms of lymph formation and composition, lymph clearance and resulting cellular effects, the impact on potential antitumour immune responses, methods to identify and measure lymphatic function, and new approaches to exploit or target lymphatics for therapy.

7.1 Introduction

Lymph forms from blood capillary exudates—the fluid that is forced out of capillaries and postcapillary venules by hydrostatic pressures, into the interstitium around the vessel. This protein-rich fluid flows through the extracellular matrix and between cells, bathing them in nutrients and oxygen, and carrying away cellular metabolites and waste products. Responding to pressure gradients within the interstitial space, openings at the junctions between adjacent lymphatic endothelial cells (LECs) drive

J. Shields (✉)
MRC Cancer Cell Unit, Hutchison/MRC Research Centre, Hills Road,
Cambridge CB2 0XZ, UK
e-mail: js970@hutchison-mrc.cam.ac.uk

L. Santambrogio (ed.), *Immunology of the Lymphatic System*,
DOI 10.1007/978-1-4614-3235-7_7, © Springer Science+Business Media New York 2013

entry of interstitial fluid, which then passes through at least one lymph node (LN). Interstitial fluid and pre-nodal lymph are essentially identical, since very little filtration occurs as a result of crossing the endothelial barrier prior to reaching the lymph nodes. Lymph that exits the efferent vessels following filtration is now a different composition. Lymph nodes filter the lymph to remove waste products, some proteins, fluid, and potential pathogens and/or pathogen components prior to its eventual return to the blood. Many afferent lymphatics from defined catchment areas drain to into the subcapsular sinus of each lymph node. Here, prevented from entering the cortex, particulates and high molecular weight molecules circulate via subcapsular and medullary sinuses before entering the efferent lymphatic and exiting the node. In contrast, lymph carrying sub-80 kDa proteins are able to pass via size exclusion pores (0.1–1 mm diameter) and percolate through narrow conduits within reticular networks (1–200 nm) (Gretz et al. 2000; Roozendaal et al. 2008; Sixt et al. 2005) before entering the medullary sinus. Both routes rely on antigen-presenting cells to screen lymph before its exit.

As with the physiological situation, interstitial fluid and lymph also form within tumors, and are collected from cancer-associated tissues. What does changes, however, is the surroundings in which lymph is generated and the tissues exposed to the resulting fluid. The environment in which lymph is formed and transported via can modify its composition having drastic effects on cells and tissues downstream. This chapter aims to address the roles of lymphatic function, lymph transport, and their far-reaching implications in cancer progression.

7.2 Lymph and the Tumor Microenvironment

Abnormal blood vessels contribute to generation of tumor interstitial fluid: Tumors require the growth of new blood vessels to cope with the increasing oxygen and metabolic demands of the rapidly increasing tumor cell mass. In contrast to the characteristic well-organized branching structures in normal microvasculature, the rapid and uncontrolled manner in which tumor-associated blood microvessels develop renders them tortuous, with abnormal branching patterns. Tumor blood vessels do not undergo normal pruning and stabilization steps such as pericyte recruitment (reviewed by McDonald and Baluk 2002). Tumor vessel endothelial cells have also been noted to be irregularly shaped and present with larger numbers of fenestrations and vesicles, loose cell–cell junctions (Hashizume et al. 2000) as well as reduced connections between endothelium and the normally tightly associated basement membrane (which is also altered in the tumor) (McDonald and Baluk 2002). The combination of these factors means that microvessels are unable to maintain their normal barrier function, and hence a feature of tumor vasculature is abnormally high vessel leakiness. Consequently, fluid and solutes are able to exit vessels much more readily than in normal cases; the accumulation of which contributes to high interstitial fluid pressures (IFP) found within tumors (Boucher et al. 1990; Heldin et al. 2004; Lunt et al. 2008; Wiig et al. 1982; Gutmann et al. 1992).

In comparison to normal tissues, where IFP is typically 0 mmHg or slightly lower (Boucher et al. 1990; Chary and Jain 1989), pressures associated with mouse mammary cancers have been measured at 2.4 mmHg in the superficial layers and rising to 23 mmHg in the center of tumors (Boucher et al. 1990; Wiig et al. 1982). In humans, this range is typically 10–40 mmHg but can rise to 60 mmHg in some tumors (Heldin et al. 2004; Lunt et al. 2008). These pressure gradients drive fluid out of the tumor into the lower pressure environment of peritumoral tissues. The movement of fluid through the tumor and associated tissues forms tumor interstitial fluid (TIF) flow. At the same time, extracellular matrix deposition and remodeling within and around the tumor generate further physical stresses, which compound the elevation seen in IFP (Fig. 7.1).

From vessel exudates to tumor interstitial fluid and lymph formation: In normal situations, the compositions of plasma exudates (immediately after leaving the capillary) and lymph have some similarities. Analysis of the proteome via mass spectrometry and 2D page techniques has begun to shed light onto the composition of lymph compared with plasma. Both lymph and plasma contain typical plasma proteins from albumin and immunoglobulin families (Leak et al. 2004). Lymph is also highly enriched with fibrinogen fragments, enzymes, catabolic products, complement components, extracellular matrix fragments, and cellular constituents (intracellular and membrane) such as histones, mitochondria, and cytoplasmic proteins (Mittal et al. 2009). The abundance of peptide fragments in pre-nodal lymph is also thought to be important for the maintenance of immune homeostasis (Clement et al. 2010, 2011).

So, what do tumor-associated lymphatic vessels collect? How does this differ physiological states? The fluid that exits abnormal vessels into the surrounding tumor is referred to as TIF. TIF is a component of the tumor microenvironment that is relatively overlooked compared to angiogenesis and lymphangiogenesis for example, yet is likely to play an equally important role in successful establishment of the tumor niche. Unlike the harvest of normal lymph, which requires vessel cannulation, collection of TIF is technically challenging. Progress in the field has been hampered due to difficulties in (a) accessing the tumor-associated tissues or vessels, given the disorganized tissue architecture, and (b) harvesting TIF without intercellular fluid contamination or directly altering the tissues that are under investigation (methods reviewed by Wiig et al. 2010). Early measurements (Gullino et al. 1964) demonstrated that as with normal lymph, albumin and immunoglobulins form a major fraction of TIF, whereas TIF has high H^+, CO_2, and lactic acid, but low glucose, oxygen, and pH compared with plasma and normal subcutaneous fluid. These differences most likely reflect the differing metabolic needs, and hypoxic environment seen in tumor vs. normal tissue. Representative of the changes in endothelial barrier function, an accumulation of low molecular weight proteins (<25 kDa) has been recorded in TIF compared with plasma, whereas those of >25 kDa are not significantly different (Stohrer et al. 2000). Following the acceptance of the tumor microenvironment as a major influencing factor in defining tumor fate (Hanahan and Weinberg 2011), there has been a renewed interest in the TIF and pre-nodal lymph. Given that these fluids envelop highly bioactive tumor cells, many

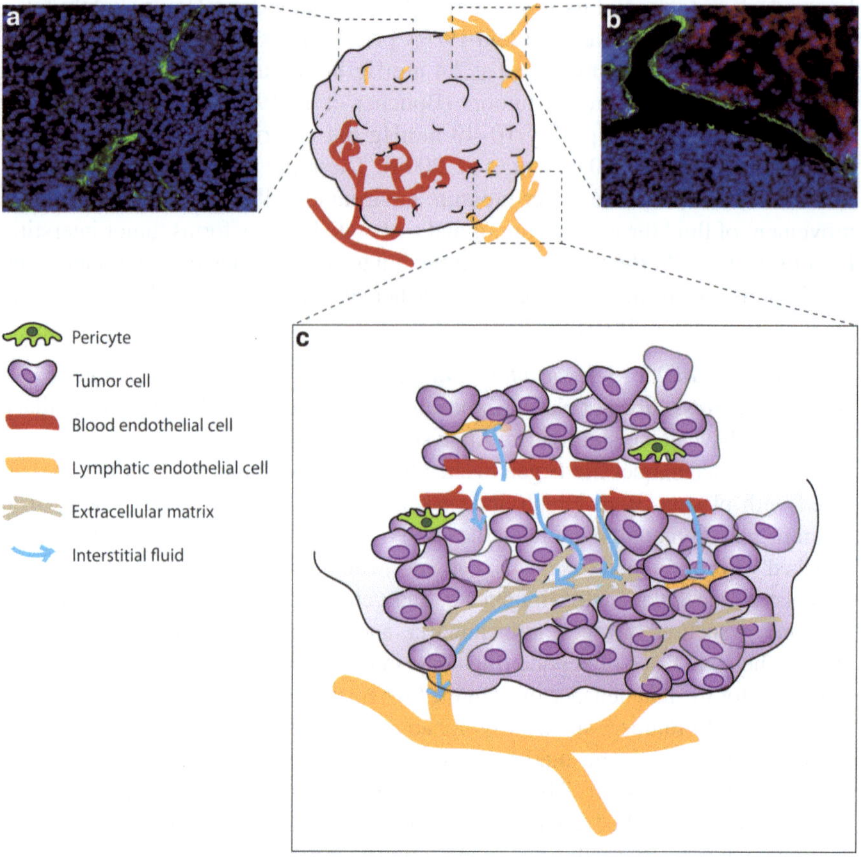

Fig. 7.1 High interstitial fluid pressure (IFP) exists within tumors. (**a**) High IFP and mechanical stresses created by high cell numbers mean that lymphatic vessels (LYVE-1 positive, *green*) within the tumor mass are collapsed and nonfunctional. (**b**) In contrast, vessels at the tumor periphery are hyperplastic and dilated. Nuclei are counterstained with DAPI. (**c**) High IFP results from exudates leaving abnormally leaky vessels within a tumor. Vessels are tortuous and associations with mural cells such as pericytes are looser than in normal tissues. As lymphatics within a tumor are not functional, fluid follows a pressure gradient via non-endothelial channels into the surrounding tissues where resistance is lower. From here, dilated peritumoral lymphatics collect the lymph and any proteins contained within

components of TIF are actively secreted or produced in metabolic exchanges, and thus represent an accurate readout of cellular events in the tumor. Breast (Celis et al. 2004, 2005; Mannello et al. 2009; Wiig et al. 2003), ovarian (Haslene-Hox et al. 2011), and renal cell (Teng et al. 2010) carcinomas are among the first to undergo proteomic profiling of TIF composition. These studies have re-highlighted the abundance of proteins contained within TIF, and identified enriched compounds with potential therapeutic significance. Of those examined, particular enrichment of plasma membrane-associated or predicted extracellular matrix constituents was recorded when compared with normal adjacent tissue (Teng et al. 2010). Given that TIF is a direct reflection of metabolic state and cellular activity and that a major hallmark of cancer is growth factor overexpression and independence (Hanahan and Weinberg 2011), we can expect that TIF is enriched with numerous other factors in addition to those described above. TIF that passes over tumor cells and surrounding tissues before entering vessels as lymph is also rich in all the growth factors, cytokines, and chemokines secreted from within the tumor microenvironment—irrespective of whether this is from tumor cells, fibroblasts, endothelial cells, or infiltrating immune cells. TIF therefore constitutes a potentially rich hunting ground for tumor-specific biomarker discovery, in both shed (e.g., components released during cell death) and soluble secreted forms (Leak et al. 2004; Clement et al. 2010, 2011; Celis et al. 2004; Haslene-Hox et al. 2011).

7.3 Physical Effects Created as a Result of Tumor-Associated Lymphatic Drainage and Lymph Flow

We have seen that TIF and lymph compositions differ from each other in physiology and disease, but what impact does this have on tumor cells themselves, their immediate surrounding tissues, and those further downstream? This section of the chapter will discuss their role in the context of lymph-mediated effects, as well as the resultant biophysical factors that contribute to the tumor microenvironment.

Tumor-associated lymphatic vessels: Within tumors, the elevated IFP from fluid and protein accumulation, along with the mechanical stresses imposed by proliferating tumor cells within a confined space results in the compression of intratumoral lymphatic vessels, rendering them nonfunctional (Leu et al. 2000; Jain and Fenton 2002; Padera et al. 2002). The loss of intratumoral lymphatic functionality and reduction in fluid clearance from the tumor interstitium compound the problem of high fluid pressure within the tumor. Instead, fluid is able to move through a tumor via non-endothelial, matrix-rich channels oozing out into surrounding tissues that pose the least resistance (Padera et al. 2002). Here, TIF can be collected by peritumoral lymphatics. In many tumor types, the tumor periphery contains an abundance of lymphatic vessels (Mandriota et al. 2001; Shields et al. 2004; Skobe et al. 2001). These vessels are either co-opted preexisting vasculature or newly formed and remodeled vessels. Peritumoral lymphatics are frequently hyperplastic and functionally abnormal having malformed valves and retrograde flow (Hagendoorn et al. 2006; Isaka et al. 2004). That being said, even with abnormal function, tumor-associated lymphatics collect the protein-rich fluids exiting the tumor mass (Fig. 7.2).

Many human cancers metastasize via the lymphatic system. To maximize the opportunity for dissemination, tumor-derived expression and secretion of vascular endothelial growth factor-C and -D (VEGF-C and VEGF-D, the major lymphangiogenic growth factors) are frequently observed in tumor cells (Mandriota et al. 2001; Shields et al. 2004; Skobe et al. 2001; Stacker et al. 2001; Mattila et al. 2002) and

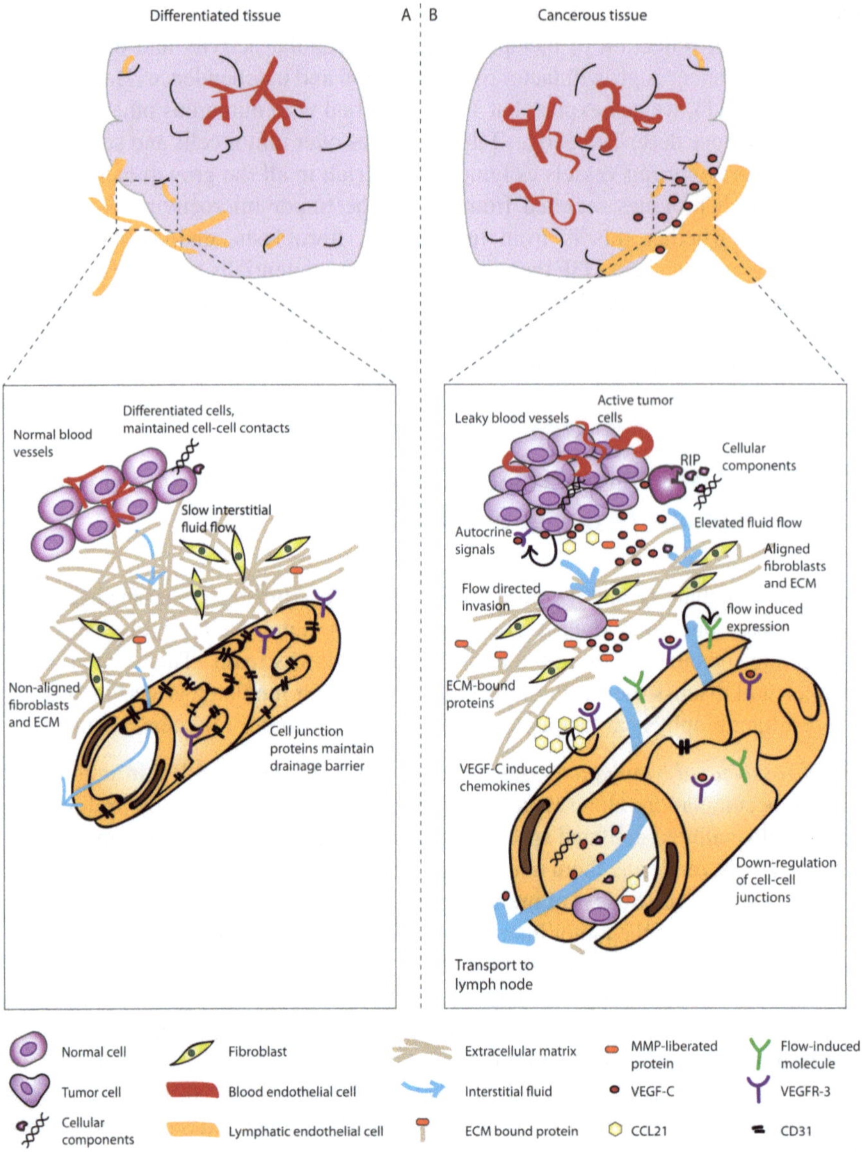

Fig. 7.2 The effects of tumor interstitial flow on tumor-associated tissues. (**a**) To ensure the maintenance of tissue homeostasis in normal differentiated tissues, plasma exudates leaving blood vessels are slowly transported through the interstitial space via pressure gradients before entering lymphatic capillaries. Cell adhesion molecules such as CD31 maintain lymphatic endothelial cell (LEC) junctions. Self-antigens and potential pathogenic components are also transported to help maintain immune homeostasis. (**b**) In the tumor microenvironment, enhanced interstitial flow created by high IFP exits the tumor and drains into peritumoral lymphatics. In the process of passing over metabolically active tumor cells, TIF and later lymph pick up growth factors (e.g., vascular endothelial growth factor-C [VEGF-C]) and tumor cell components, which enter the lymphatic

surrounding tissues (Gallego et al. 2011; Schoppmann et al. 2006). VEGF-C expression in tumor-associated tissues reflects a strong correlation with the incidence of lymph node metastasis and poor prognosis in patients (Emmett et al. 2010; Kurahara et al. 2010; Nakamura et al. 2005). VEGF-C directly acts on lymphatic vessels at the tumor periphery through the ligation of its cognate receptor VEGFR-3, stimulating both de novo vessel formation and remodeling (hyper-proliferation) of those preexisting. The expanded lymphatic network presents a tumor with greater chance of encountering a vessel, and a portal for tumor metastasis that can directly help tumor progression.

Increased lymphatic drainage: The stimulation of extra lymphatic vessels also confers indirect effects that help support a growing tumor. TIF that oozes out of a tumor is collected as lymph by peritumoral lymphatics. VEGF-C (and other growth factors such as FGF) present in the vicinity of the tumor is sufficient to induce high peritumoral vessel density, so although hyperplastic and functionally abnormal, the increase in number is translated to increased functional output and enhanced capacity for fluid clearance (Hoshida et al. 2006). Moreover, VEGF-D has been shown to act further downstream in the vessel hierarchy via a mechanism distinct from VEGF-C. Rather than stimulating endothelial proliferation, VEGF-D induces collecting lymphatic vessel dilation in a prostaglandin-dependent manner (Karnezis et al. 2012). This would imply that the synergies between tumor-derived VEGF-C and -D extend beyond the induction of new lymphatic vessels, towards modulating their functionality in the tumors' favor.

Fluid flux effects on cell behavior: The physical movement of fluid through tumor tissues and the tissues immediately surrounding it is a tumor-promoting stimulus in its own right. Firstly, it is likely that tumor cells in the process of detaching from the main tumor bulk are assisted as a consequence of increasing lymph flow. Shedding cells may be physically carried along fluid channels towards draining lymphatics and further downstream, augmenting delivery of "flushed" cells to the draining lymph node (Hoshida et al. 2006). In lymphatic-rich tumors, there is generally more cellular movement between the tumor and lymph nodes (Shields et al. 2010; Hoshida et al. 2006; Lund et al. 2012; Hoshida et al. 2006). Fluorescently labeled microbeads have enabled researchers to define specific migrating cell populations and trace cell movements. Not surprisingly, the major cell types identified as trafficking via the tumor lymphatics are immune cells (immune context discussed later) (Shields et al. 2010; Lund et al. 2012). In addition to physically

Fig. 7.2 (continued) system towards the draining lymph node where further changes occur in preparation for metastasis. Tumor-derived growth factors can synergize with flow to act in an autocrine fashion increasing cell invasiveness, or paracrine on cancer-associated fibroblasts and LECs. VEGF-C stimulates lymphatic growth (enhancing drainage capacity), and also the secretion of lymph homing chemokines such as CCL21. Proteases, skewed by flow, further liberate proteins within the matrix amplifying gradient effects. Flow itself is sufficient to induce environmental changes, stimulating remodeling of the interstitium and modification of vessel functionality (via down-regulation of junction proteins and changes in surface expression) assisting tumor cell detachment and metastasis, fluid convection created by lymphatic drainage can act as a morphogenetic cue, synergizing with the local chemical and physical environment. Together, these forces can significantly influence the environment and cellular behaviors. Oncogenic stresses and physical pressures associated with tumor development stimulate tumor cells to secrete growth factors, chemokines, etc.; for example, high pressures within a tumor have been shown to stimulate tumor proliferation (Hofmann et al. 2006) and the release of VEGF-C (Nathan et al. 2009). These factors exert their effects on cells within their immediate proximity in (a) autocrine, (b) paracrine fashion, or are (c) transported to downstream tissues to exert their effects remotely (Gretz et al. 2000; Helm et al. 2005; Miteva et al. 2010; Ng et al. 2004). *Autocrine effects on tumor cells*: On exposure to subtle fluid flows, transcellular chemokine gradients are generated that are biased towards functional, draining lymphatic vessels. In this sense, a tumor cell can follow an autologously generated cue that directs it to the nearest functioning vessel—and escape route (Fleury et al. 2006; Haessler et al. 2012; Shields et al. 2007a; Polacheck et al. 2011). This phenomenon, referred to as autologous chemotaxis, was shown in vitro in tumor cells that utilize a CCL21–CCR7 signaling loop, but has the potential to exist for any number of tumor-derived factors. Similarly, VEGF-C is able to ligate its receptor VEGFR-3 in an autocrine manner leading to increased tumor cell motility and proteolytic capacity (Issa et al. 2009). *Paracrine effects on cells within the stroma*: Proteins carried within the fluid that is collected from a tumor maintain the capacity to influence neighboring cells encountered en route to lymphatic vessels, including cancer-associated fibroblasts and the LECs themselves. Fibroblasts proliferate and become activated upon exposure to flow. Flow-dependent responses measured include the up-regulation of smooth muscle actin for contractility, reorientation to align themselves and collagen fibers perpendicular to flow, the induction of factors such as TGF-β_1, and enhanced degradation of surrounding matrix (Ng et al. 2005; Ng and Swartz 2003, 2006; Shieh et al. 2011). In this way, flow through tissue spaces can "prime" the tumors surroundings in preparation for metastasis. Indeed, it has been shown that activated cancer-associated fibroblasts guide escaping tumor cells away from the primary mass towards vessels (Shieh et al. 2011; Gaggioli et al. 2007). We also know that the extracellular matrix which fibroblasts deposit modulates the physical environment via the mechanical stresses they exert, but matrix components can also amplify the paracrine (and some degree autocrine) effects generated by interstitial fluid flow. Many proteins bind to matrix components leading to sequestered pools of locally high concentrations of e.g. latent TGF-β_1, VEGF, or CCL21 (Shieh et al. 2011; Patel et al. 2001; Zilberberg et al. 2012). Proteases transported within the interstitial fluid are able to rapidly liberate any bound factors. Directional fluid flow and protein release can then synergize to form directed and intensified chemical gradients that cancer-associated fibroblasts or egressing tumor and immune cells can respond to (Ng et al. 2004, 2005; Haessler et al. 2012; Shields et al. 2007a, b; Polacheck et al. 2011). Movement of lymph as it enters lymphatics (referred to as transmural flow) and lymph constituents modulates LEC properties too. The lymph node homing chemokine CCL21 is produced and secreted in response to transmural flow, acting as a potential homing signal for CCR7-expressing tumor cells in the early stages of metastasis. CCL21 can in turn simulate VEGF-C up-regulation, which as we saw earlier may further contribute to lymphatic mediated events. Furthermore, even at low flow rates of 0.1 μm s^{-1} (representative of measured interstitial fluid velocity), entry of lymph into vessels results in delocalization and down-regulation of cell junction proteins such as VE-cadherin and CD31 (Miteva et al. 2010), but up-regulation of cell adhesion molecules such as ICAM and E-selectin; a preparatory step prior to cellular transmigration and entry into lymphatic circulation. *Downstream effects within vessels and lymph node*: Once inside a vessel, tumor cell behavior continues to be influenced by lymph flow and lymph-borne factors. When compared with the harsh environment of the blood system, lymph flows seen in tumor-draining lymphatics provide a setting more conducive with cell survival—with velocities typically an order of magnitude lower than blood (1–10 vs. 100–1,000 μm s^{-1}, respectively; Berk et al. 1996; Leu et al. 1994a, b; Swartz et al. 1996). Tumor emboli within vessels have been shown to utilize the low shear stresses found in lymphatics to their advantage. Under low flow conditions ($\tau = 2.5$ dyn cm^{-2}; Byers et al. 1995), highly invasive E-cadherin negative or defective tumor cells are much more likely to detach from their counterparts in response to lymph flow and lodge in local lymph nodes than E-cadherin positive epithelial cells (Byers et al. 1995).

Lymph drainage from the tumor site clears the excess fluid from leaky blood vessels, which as we discussed can have autocrine and paracrine effects. As lymph flows along a vessel hierarchy, its components also pass this same route and therefore imply that tumors can utilize this path to remotely access distant tissues. One of the primary reasons a tumor might want to communicate with distant tissues is in order to prepare the new environment for metastasizing cells, otherwise known as the pre-metastatic niche (reviewed by Peinado et al. 2011; Psaila and Lyden 2009). Changes to lymph nodes immediately downstream of tumor-bearing sites *prior* to arrival of metastasizing cells have frequently been recorded in experimental models (Harrell et al. 2007; Hirakawa et al. 2005, 2007; Mumprecht et al. 2010; Ruddell et al. 2008), and similar changes in clinical samples are showing prognostic potential (Kurahara et al. 2010; Jakob et al. 2011). In mouse models of lymph node metastasis, tumor-draining lymph nodes present with greatly enlarged and supernumerary lymphatic sinuses before metastasis occurs, whereas lymph node metastases are not observed in the absence of LN lymphatic expansion (Hirakawa et al. 2007). In contrast, metastasis is frequently observed where LN lymphangiogenesis had preceded, and further distant lesions, e.g., lung, are more likely to occur (Hirakawa et al. 2007). Whether this directly relates to an enhanced delivery of tumor cells to the node, or some secondary survival advantage provided via environmental adaptation is still not clear. Consistent with earlier data on VEGF-C effects, lymph flow to the nodes is enhanced suggesting that tumor-derived VEGF-C not only impacts tumor cells and LECs at the tumor periphery but that VEGF-C can also be transported in lymph to the LN to stimulate lymphangiogenesis there.

7.4 Lymph Effects on the Tumor Immune Response

Of course, there is a flip side to every story. In the case of lymph and lymphatics from the context of a tumor, we know that tumor-derived factors and proteins contained within lymph can help to increase the probability of a tumor cell finding lymphatic vessels. We have also seen that the lymph flow generated as a consequence can both physically detach cells and act as a guidance cue. However, in providing a route for tumor cell dissemination, lymphatic expansion also reinforces the connection to our immune system. To survive, therefore, a tumor must develop methods to suppress potentially destructive immune responses. Early clues alluding to a cooperative relationship between a tumor, lymphatics, and the host immune system came from studies illustrating that immune cells recruited to a tumor and within the draining LN actively stimulated lymphangiogenesis and tumor progression (Schoppmann et al. 2006; Harrell et al. 2007; Jeon et al. 2008; Moussai et al. 2011). More recently, research advances indicate that tumors are capable of exploiting normal aspects of immunology into their advantage, confirming that lymph and lymphatic vessels are important mediators of this. Maintenance of tissue fluid balance requires an equilibrium between fluid that enters the interstitial space and that which leaves via the lymphatics. This ensures a flow of lymph from the periphery via lymph nodes and towards the thoracic duct before reentering blood circulation. The anatomical sites of lymphatic capillaries in peripheral tissues mean that they are in close proximity to the environment, and therefore represent a major route of immune surveillance and protection. In addition to plasma proteins and cellular waste products lymph also acts as a sampling reservoir, transporting soluble antigens from peripheral tissues towards the LN (Clement et al. 2010; Sixt et al. 2005; Roozendaal et al. 2009). Here, antigen and small molecular weight proteins penetrate the lymph node to the T cell zones via fine conduit structures and lymph node-resident antigen-presenting cells sample the unfiltered fluid, take up, and cross present antigen. These responses are rapid (within minutes), much more so than mobilizing peripheral antigen-presenting cells (8–12 h). Delivery of soluble antigens represents a rapid efficient way to monitor events at the periphery and adapt accordingly, either by inciting effective immune responses or by preventing inappropriate tissue-damaging responses. Many tumors evolve to express mutated antigens and overexpress normal tissue antigens (e.g., EGFR), differentiation antigens specific to tissue types (e.g., Melan-A specific to skin), or cancer/testes antigens, normally only expressed in germline tissues. In normal circumstances such changes would be detected and eliminated, however this is not the case in cancer. Something clearly goes awry. With collection of tumor-derived factors, vesicles, chemokines, inflammatory mediators, and cells by lymphatics it is also likely that shed tumor-derived antigens are carried in the soup. Influenced by this altered milieu, normally reactive immune cells undergo changes to functional phenotype (e.g., induction of T_{reg}), functional inactivation (loss of co-stimulatory molecules in dendritic cells), or die (deletion and exhaustion of tumor-reactive T cells). Indeed in physiological situations, lymph node LECs and stromal cells, which are continually exposed to lymph-borne antigens, are able to process and present endogenous antigen, resulting in the deletion of self-reactive T cells and maintenance of immune homeostasis and tolerance (Lee et al. 2007; Cohen et al. 2010; Fletcher et al. 2010). More recently, this scenario was translated to the pathological setting of the tumor. Tumors that express the model antigen ovalbumin together with VEGF-C developed extensive peritumoral lymphatic vessels networks, significant immune infiltrates which concentrated in close proximity to the vessels, and enhanced lymphatic drainage (Lund et al. 2012). This study demonstrated that tumor-specific antigen could be scavenged by LECs of the tumor and draining LN and presented in Class I MHC complexes. This leads to impaired activation, function, and deletion of ovalbumin-specific, tumor-reactive T cells that infiltrate the tumor (Lund et al. 2012). Consistent with this, lymph node LECs are capable of iNOS production upon inflammatory stimulation, which attenuate T cell proliferation (Lukacs-Kornek et al. 2011) and may promote regulatory T cell development (Niedbala et al. 2007). In areas of chronic inflammation such as the tumor microenvironment, the associated immune cells can further contribute to immune suppressive effects by signaling to collecting lymphatics in their vicinity. iNOS secreting immune cells such as myeloid-derived suppressor cells (MDSC) cause collecting vessel relaxation and diminished vessel contraction strength (Liao et al. 2011). While the interpretations of these and other studies differ, they all support the hypothesis that tumors employ complex strategies to manipulate our immune system, but thanks to the major role that lymphatics and their cargo play in immune homeostasis, modifications to lymphatic function are a key weapon in a tumors arsenal.

7.5 Lymph Measurement in Tumors

The steps leading to hematogenous metastasis have been extensively studied (Butler and Gullino 1975; Condeelis and Segall 2003; Liotta et al. 1974) and until recently, lymphatics had not received the same attention. However, genetically modified mouse models are now proving invaluable tools to aid our understanding of how lymphatic vessels function in both normal and diseased tissues.

Detection of lymphatic vessels: Coincident improvements to intravital imaging techniques and development of transgenic lines with fluorescent proteins such as GFP, mOrange, tomato red, and luciferase under the transcriptional control of lymphatic-specific genes *Prox1* or *Vegfr3* have made the accurate identification of lymphatic vessels possible and allowed real-time, minimally invasive tracking of the cells that contribute to the changes (Choi et al. 2011; Hagerling et al. 2011; Martinez-Corral et al. 2012; Truman et al. 2012). This is particularly useful for studying the onset and progression of lymphangiogenesis in cancer both at primary and metastatic sites (Harrell et al. 2007; Mumprecht et al. 2010; Martinez-Corral et al. 2012).

Measuring functionality in tumors: As we have seen, vessel functionality is essential for the establishment and propagation of tumors. Using murine models of cancer, lymph exit from tumors and transit to lymph nodes have been measured by magnetic resonance imaging (MRI) technology (Ruddell et al. 2008; Dafni et al. 2002) and fluorescence microlymphangiography using fluorescently labeled tracers (dextrans and quantum dots) (Hoshida et al. 2006; Harrell et al. 2007; and reviewed by Cohen et al. 2011). Both methods have advantages and disadvantages. MRI allows three-dimensional imaging, which can be achieved for both superficial and deeper structures (Pathak et al. 2005). Contrast agents such as low molecular weight gadolinium-based agents (e.g., dimeglumine gadopentate, Gd-DTPA), however, can rapidly diffuse out of lymphatic vessels so are not suitable for long-term imaging in small animal models. Gd-DTPA has proven useful for short-term kinetic studies of tumor-associated lymph flow in mice, where rapid uptake into lymphatics enabled measurement of the increased lymph flow from tumors to lymph nodes and back into the circulation of tumor-bearing mice (Ruddell et al. 2008). In contrast, quantification of lymph flow velocity is achieved using fluorescence lymphangiography, which tracks the convective movement of photobleached spots within fluorescently loaded vessels (Hoshida et al. 2006). Alternative methods to quantify lymphatic function involve the measurement of depot clearance of labeled albumins or dextrans (Emmett et al. 2011; Karlsen et al. 2012). Fluorescent and near-infrared optical imaging are useful for imaging of superficial lymphatic vessel drainage, however, they do have the drawbacks that images are of relatively low resolution and deeper vessels cannot be detected (Harrell et al. 2007; Kwon and Sevick-Muraca 2007). Recently, optical frequency domain imaging, a second generation of optical coherence tomography, has been described as an exciting new technique for intravital imaging of tumors, utilizing elastic light scattering properties to yield high resolution images in 3D. Signals are generated based upon the intrinsic movement of erythrocytes so no contrast agents are required. Without the need for tracers, both angiography and lymphangiography can be employed at the same time and distinguished based on contrast. This method is being advocated as a high volumetric imaging technique with sufficiently high resolution and offering enhanced tissue penetration up to depths in the order of millimeters (Vakoc et al. 2009).

Lymph flow measurements in cancer patients: In patients, tracers that rely on the functional properties of lymphatics for selective uptake and transport of either radiolabeled/radio-opaque colloids or inert dyes are most commonly used to locate sentinel lymph nodes (Lai and Rockall 2010; Mouli et al. 2010). There are few studies that have measured lymph flow in cancer patients. Indocyanine green, a near-infrared fluorophore, has recently been used to measure lymph velocities, and lymphatic functionality in breast cancer patients undergoing sentinel lymph node mapping (Rasmussen et al. 2009; Sevick-Muraca et al. 2008). This noninvasive technique measured lymph velocities in patients ranging from 0.08 to 0.32 cm s^{-1} (4.8–19.2 cm min^{-1}) (Sevick-Muraca et al. 2008) in contrast to normal arms where lymph velocities have been measured at 8.9 cm min^{-1} (Modi et al. 2007a, b). Near-infrared fluorescence and other imaging techniques are also being utilized to map vessel functionality in patients suffering from complications associated with surgical resection of lymph nodes. These modalities improve the efficiency of manual lymph drainage therapy (Maus et al. 2012; Sevick-Muraca 2012), and can also be applied to help detect early changes or deficits in vessel function that could account for lymphatic dysfunction preceding postmastectomy lymphedema (Modi et al. 2007a; de Rezende et al. 2011; Stanton et al. 2006; Szuba et al. 2007).

7.6 Lymph and Lymphatics as Therapeutic Targets

With increasing insight into the role of lymphatics, their function, and the compounds they carry, comes an increasing aspiration to exploit or manipulate these features as therapeutic targets for the treatment of cancer (Fig. 7.3).

Manipulation of VEGF-C/VEGFR signaling axis: Numerous studies have been conducted with a goal to assess the therapeutic potential of antibodies raised against lymphatic-specific epitopes. In experimental cancer models, neutralizing antibodies and antibody fragments against VEGF-C and its receptors inhibit the growth of peritumoral lymphatics and ultimately the incidence of lymph node metastasis but have no effect on growth of the primary tumor (Rinderknecht et al. 2010; Roberts et al. 2006; Tvorogov et al. 2010; Yang et al. 2011). Major effects on the primary tumor and metastasis were only observed when combinations of both anti-VEGFR-2 and -VEGFR-3 were used. Moreover, antibodies designed to block shared markers of blood and lymphatic endothelium such as ephrins and angiopoietins impact both systems but the mechanisms of actions are yet unclear, whether they exert their beneficial effects on lymphatics due to direct inhibition, or whether the observed lymphatic inhibition is a downstream consequence of the forced reduction in blood vessel density within the tumor (Abengozar et al. 2012; Holopainen et al. 2012; Hwang-Bo et al. 2012). Small molecule inhibitors such as Sunitinib, Sorafineb, PTK787/ZK222584, and E7080, and mTOR inhibitors such as Rapamycin have also been shown to impact tumor lymphangiogenesis and lymphatic dissemination (Kodera et al. 2011; Matsui et al. 2008; Patel et al. 2011; Schomber et al. 2009). While effective in preclinical models, these compounds, e.g., broad-spectrum tyrosine kinase inhibitors, are not lymphatic-specific, therefore off-target secondary effects need to be carefully evaluated. These data would suggest that with the complex survival strategies put in place by a growing tumor, targeting either lymphatics or blood vessels alone is not the best way to attack tumors.

Particulates that exploit the lymphatics: As we learn more about how lymphatics function and the proficient manner in which lymphatics are able to rapidly but selectively take up pathogens such as viruses and virus-like particles up to 150 nm in diameter, there is an increasing thrust to harness and exploit these properties. With this in mind, researchers are aiming to improve the efficacy of drug delivery, and develop novel "vaccines" that target both the tumor and the downstream immune checkpoints bridged by the lymphatics. Compounds loaded within particulates that are drained and carried via

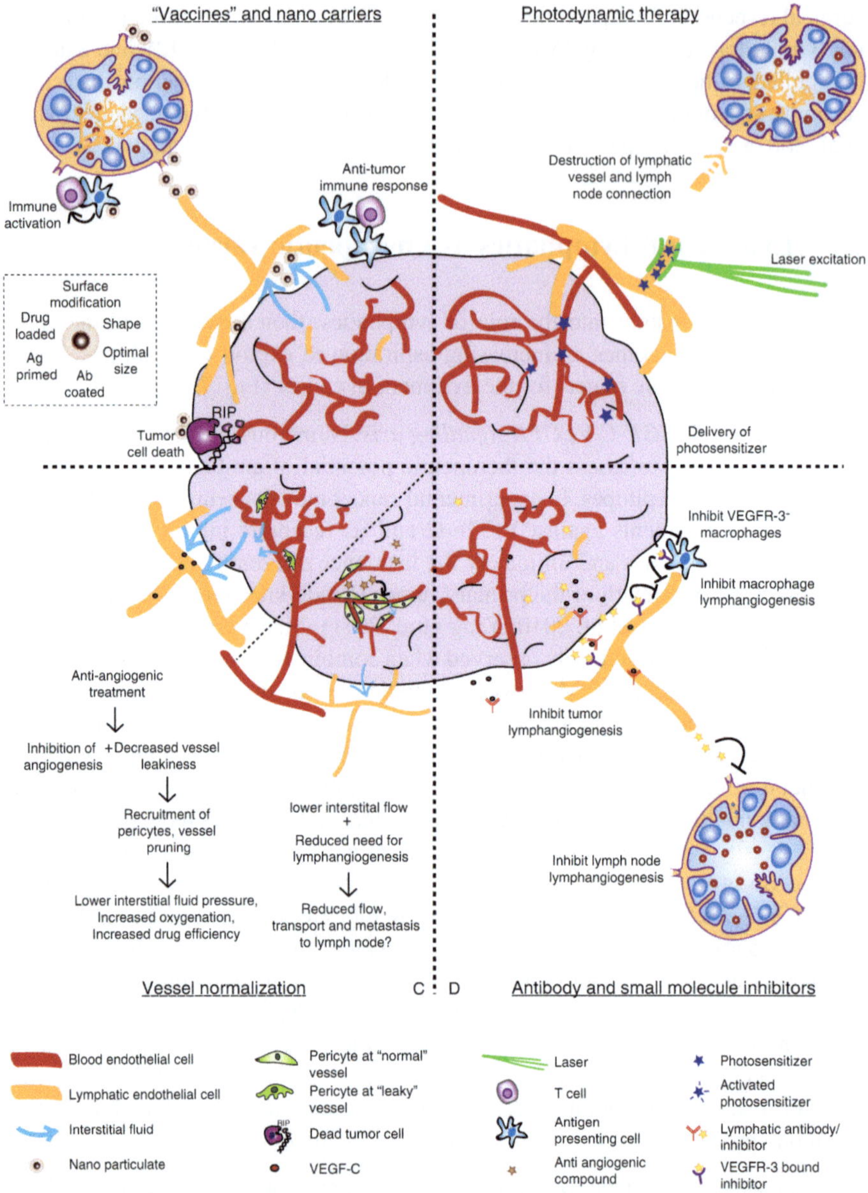

Fig. 7.3 Lymph-associated therapeutic strategies. (**a**) The implementation of "vaccines" and nanocarriers aim to improve efficiency of therapies by exploiting the properties of lymph and their transport via lymphatics. These particulates can come in many guises due to their pliability. Researchers hope to use this strategy as an immunotherapy, directly targeting the draining lymph node and immune system, to more effectively deliver cytotoxic payloads or kill tumor cells as a direct consequence of particle properties. (**b**) Photodynamic therapy (PDT) relies on the accumulation of photosensitizers within tissues, but cytotoxic activation only occurs once the specific tissue is exposed to light of defined wavelengths. Preclinical models demonstrated this method could

the lymphatics display reduced nonspecific cytotoxicity and enhanced bioavailability (e.g., protection from enzymatic degradation and first pass clearance from liver metabolism) compared with intravenous delivery. Furthermore, the pliability of such particulates, e.g., particle material, core composition, size, structure, shape, and surface modification adds to their attractiveness as a therapeutic vehicle. For example, the potential to modify their surface chemistry can enhance lymphatic uptake (sub-50 nm diameter), retention (avidin coating), and cell-specific uptake. Ingestion particularly by antigen-presenting cells therefore potentiates these nanoparticles as immune activators. Although the use of patient-specific therapeutic vaccines is still some way off, the prospect of individualized lymph-targeted immune-modulatory therapies is immensely exciting. The number of scientific research groups, biotech start-ups, and pharmaceutical giants buying into this technology indicates that the notion of lymph-borne therapeutics is in fact likely to become a powerful therapeutic platform.

Photodynamic ablation therapy: With the knowledge that systemic delivery of blocking antibodies is less effective and specific than hypothesized, alternative and more targeted ideas are under exploration. Photodynamic therapy (PDT) typically works via the accumulation of compounds with photosensitizing properties at your tissue of choice. Exposure of this area to visible light activates the photosensitizer generating reactive oxygen species and cytotoxicity tissues exposed to the compound (reviewed by Juarranz et al. 2008). The benzoporphyrin derivative verteporfin is clinically applied to patients with diseases such as age related macular degeneration (Zhao et al. 2010). It is delivered to patients in the form of liposomes. Recently, taking advantage of lymphatic functional characteristics, these particulates were applied to experimental models of melanoma. Upon intradermal inoculation, liposomes in the tumor-associated interstitium drained into local lymphatics where they were retained for up to 2 h. Upon subsequent activation by 639 nm laser light, activation of the verteporfin resulted in specific destruction of tumor-associated lymphatics but also cancer cells contained within them in transit to the lymph node (Tammela et al. 2011). Developments of this nature are particularly exciting because of their low toxicity and off-target effects when compared to conventional cytotoxic therapies. Further advantages offered by PDT in the context of tumor-associated lymph formation are that this therapy can also (a) induce vascular shutdown which may help to interrupt the nutrient supply chain but also switch off vascular leakage and high interstitial pressures that drive lymph formation; and (b) activate the immune response

Fig. 7.3 (continued) effectively ablate tumor-draining lymphatics and cancer cells within them. How this would impact tissue fluid balance, however, is not clear. (**c**) Vessel normalization strategies use anti-angiogenic therapies to temporarily restore the balance of pro- and anti-angiogenic signals, thereby reducing vessel leakiness and fluid pressure, improving oxygenation and drug delivery. Reduction of fluid pressure may then decrease lymph flow and the effects it and its cargo impart on lymphatics and downstream tissues. (**d**) Anti-lymphatic antibodies and small molecule inhibitors may prevent VEGFR-3 signaling, and therefore lymphangiogenesis at the tumor and lymph node by blocking the receptor or scavenging ligands on responsive cells (e.g. endothelium and infiltrating immune cells). Secondary advantageous effects may also occur via interactions with blood endothelial cells

which is compromised in tumors, partly as a consequence of the immune modulating components transported within tumor-derived lymph.

Vessel normalization strategies: Rather than directly targeting lymph and lymphatic vessels, the route of *vessel normalization* is also being explored to target abnormal vessels within the tumor. The concept came from the observation that in preclinical models, treatment with anti-angiogenic agents transiently restored vessel function in tumors, improving oxygenation and reducing fluid pressure (Tong et al. 2004; Winkler et al. 2004; Yuan et al. 1996). In this window of opportunity, the efficacy of cytotoxic drugs was also improved (Winkler et al. 2004; Mazzone et al. 2009). The reinstatement of a more normal vessel network by balancing pro- and anti-angiogenic signals within the tumor microenvironment may therefore help to reduce the manifestation of lymph-mediated disease downstream stemming from high TIF pressure, irrespective of whether this occurs through direct effects on vessel wall components or via inhibition of angiogenic myeloid cells (reviewed by Carmeliet and Jain 2011). Direct evidence to the benefits of this concept in human tumors is still missing, however recent studies in colorectal cancer and glioblastoma have reported changes in Bevacizumab-treated patients that are consistent with the notion of normalization (Batchelor et al. 2007, 2010; Willett et al. 2004). Although in theory, vessel normalization may attenuate lymphatic promotion of tumor dissemination by reducing TIF generation and flow thereby reducing need for lymphangiogenesis and decreasing transport of soluble pro-tumor factors, recent work has demonstrated that caution is indeed warranted: compounds with potential to induce vessel normalization (and increase drug delivery) may also promote lymphatic metastasis (Grepin et al. 2012; Liu et al. 2011). Therefore understanding the delicate balances that exist within the tumor environment will be critical when designing and optimizing the doses, combinations, or timings of future therapeutic platforms.

7.7 Concluding Remarks

There is no doubt that lymph via its generation, transport, and the vessels that facilitate this plays an essential role in the pathophysiology of cancer, and carries widespread implications. As a result, lymphatics and their function are rapidly becoming a major therapeutic target. It is also clear to see that the relationship between tumor blood vessels, lymphatics, and their microenvironment is a convoluted one with multiple levels of complexity. But as the field grows and expands into previously unexplored niches, we have begun to embrace new methodologies. The integration of tumor biology with disciplines, such as physics, proteomics, and bioengineering for example, is helping to unravel these relationships layer by layer at continually evolving and accelerating rates. It is through collaborations such as these that the biggest breakthroughs will transpire; an event we await with much anticipation.

References

Abengozar MA et al (2012) Blocking ephrin-B2 with highly specific antibodies inhibits angiogenesis, lymphangiogenesis, and tumor growth. Blood 119:4565–4576

Batchelor TT et al (2007) AZD2171, a pan-VEGF receptor tyrosine kinase inhibitor, normalizes tumor vasculature and alleviates edema in glioblastoma patients. Cancer Cell 11:83–95

Batchelor TT et al (2010) Phase II study of cediranib, an oral pan-vascular endothelial growth factor receptor tyrosine kinase inhibitor, in patients with recurrent glioblastoma. J Clin Oncol 28:2817–2823

Berk DA, Swartz MA, Leu AJ, Jain RK (1996) Transport in lymphatic capillaries. II. Microscopic velocity measurement with fluorescence photobleaching. Am J Physiol 270:H330–H337

Boucher Y, Baxter LT, Jain RK (1990) Interstitial pressure gradients in tissue-isolated and subcutaneous tumors: implications for therapy. Cancer Res 50:4478–4484

Butler TP, Gullino PM (1975) Quantitation of cell shedding into efferent blood of mammary adenocarcinoma. Cancer Res 35:512–516

Byers SW, Sommers CL, Hoxter B, Mercurio AM, Tozeren A (1995) Role of E-cadherin in the response of tumor cell aggregates to lymphatic, venous and arterial flow: measurement of cell-cell adhesion strength. J Cell Sci 108(Pt 5):2053–2064

Carmeliet P, Jain RK (2011) Principles and mechanisms of vessel normalization for cancer and other angiogenic diseases. Nat Rev Drug Discov 10:417–427

Celis JE et al (2004) Proteomic characterization of the interstitial fluid perfusing the breast tumor microenvironment: a novel resource for biomarker and therapeutic target discovery. Mol Cell Proteomics 3:327–344

Celis JE et al (2005) Identification of extracellular and intracellular signaling components of the mammary adipose tissue and its interstitial fluid in high risk breast cancer patients: toward dissecting the molecular circuitry of epithelial-adipocyte stromal cell interactions. Mol Cell Proteomics 4:492–522

Chary SR, Jain RK (1989) Direct measurement of interstitial convection and diffusion of albumin in normal and neoplastic tissues by fluorescence photobleaching. Proc Natl Acad Sci USA 86:5385–5389

Choi I et al (2011) Visualization of lymphatic vessels by Prox1-promoter directed GFP reporter in a bacterial artificial chromosome-based transgenic mouse. Blood 117:362–365

Clement CC et al (2010) An expanded self-antigen peptidome is carried by the human lymph as compared to the plasma. PLoS One 5:e9863

Clement CC, Rotzschke O, Santambrogio L (2011) The lymph as a pool of self-antigens. Trends Immunol 32:6–11

Cohen JN et al (2010) Lymph node-resident lymphatic endothelial cells mediate peripheral tolerance via Aire-independent direct antigen presentation. J Exp Med 207:681–688

Cohen SM, Fishinghawk BG, Cohen MS (2011) Translational imaging of lymphatics in cancer. Adv Drug Deliv Rev 63:956–962

Condeelis J, Segall JE (2003) Intravital imaging of cell movement in tumours. Nat Rev Cancer 3:921–930

Dafni H, Israely T, Bhujwalla ZM, Benjamin LE, Neeman M (2002) Overexpression of vascular endothelial growth factor 165 drives peritumor interstitial convection and induces lymphatic drain: magnetic resonance imaging, confocal microscopy, and histological tracking of triple-labeled albumin. Cancer Res 62:6731–6739

de Rezende LF, Pedras FV, Ramos CD, Gurgel MS (2011) Evaluation of lymphatic compensation by lymphoscintigraphy in the postoperative period of breast cancer surgery with axillary dissection. Tumori 97:309–315

Emmett MS et al (2010) Prediction of melanoma metastasis by the Shields index based on lymphatic vessel density. BMC Cancer 10:208

Emmett MS, Lanati S, Dunn DB, Stone OA, Bates DO (2011) CCR7 mediates directed growth of melanomas towards lymphatics. Microcirculation 18:172–182

Fletcher AL et al (2010) Lymph node fibroblastic reticular cells directly present peripheral tissue antigen under steady-state and inflammatory conditions. J Exp Med 207:689–697
Fleury ME, Boardman KC, Swartz MA (2006) Autologous morphogen gradients by subtle interstitial flow and matrix interactions. Biophys J 91:113–121
Gaggioli C et al (2007) Fibroblast-led collective invasion of carcinoma cells with differing roles for RhoGTPases in leading and following cells. Nat Cell Biol 9:1392–1400
Gallego E et al (2011) Stromal expression of vascular endothelial growth factor C is relevant to predict sentinel lymph node status in melanomas. Virchows Arch 458:621–630
Grepin R et al (2012) Acceleration of clear cell renal cell carcinoma growth in mice following bevacizumab/Avastin treatment: the role of CXCL cytokines. Oncogene 31:1683–1694
Gretz JE, Norbury CC, Anderson AO, Proudfoot AE, Shaw S (2000) Lymph-borne chemokines and other low molecular weight molecules reach high endothelial venules via specialized conduits while a functional barrier limits access to the lymphocyte microenvironments in lymph node cortex. J Exp Med 192:1425–1440
Gullino PM, Clark SH, Grantham FH (1964) The interstitial fluid of solid tumors. Cancer Res 24:780–794
Gutmann R et al (1992) Interstitial hypertension in head and neck tumors in patients: correlation with tumor size. Cancer Res 52:1993–1995
Haessler U, Teo JC, Foretay D, Renaud P, Swartz MA (2012) Migration dynamics of breast cancer cells in a tunable 3D interstitial flow chamber. Integr Biol (Camb) 4:401–409
Hagendoorn J et al (2006) Onset of abnormal blood and lymphatic vessel function and interstitial hypertension in early stages of carcinogenesis. Cancer Res 66:3360–3364
Hagerling R, Pollmann C, Kremer L, Andresen V, Kiefer F (2011) Intravital two-photon microscopy of lymphatic vessel development and function using a transgenic Prox1 promoter-directed mOrange2 reporter mouse. Biochem Soc Trans 39:1674–1681
Hanahan D, Weinberg RA (2011) Hallmarks of cancer: the next generation. Cell 144:646–674
Harrell M, Iritani B, Ruddell A (2007) Tumor-induced sentinel lymph node lymphangiogenesis and increased lymph flow precede melanoma metastasis. Am J Pathol 170:774–786
Hashizume H et al (2000) Openings between defective endothelial cells explain tumor vessel leakiness. Am J Pathol 156:1363–1380
Haslene-Hox H et al (2011) A new method for isolation of interstitial fluid from human solid tumors applied to proteomic analysis of ovarian carcinoma tissue. PLoS One 6:e19217
Heldin CH, Rubin K, Pietras K, Ostman A (2004) High interstitial fluid pressure—an obstacle in cancer therapy. Nat Rev Cancer 4:806–813
Helm CL, Fleury ME, Zisch AH, Boschetti F, Swartz MA (2005) Synergy between interstitial flow and VEGF directs capillary morphogenesis in vitro through a gradient amplification mechanism. Proc Natl Acad Sci USA 102:15779–15784
Hirakawa S et al (2005) VEGF-A induces tumor and sentinel lymph node lymphangiogenesis and promotes lymphatic metastasis. J Exp Med 201:1089–1099
Hirakawa S et al (2007) VEGF-C-induced lymphangiogenesis in sentinel lymph nodes promotes tumor metastasis to distant sites. Blood 109:1010–1017
Hofmann M et al (2006) Lowering of tumor interstitial fluid pressure reduces tumor cell proliferation in a xenograft tumor model. Neoplasia 8:89–95
Holopainen T et al (2012) Effects of angiopoietin-2-blocking antibody on endothelial cell-cell junctions and lung metastasis. J Natl Cancer Inst 104:461–475
Hoshida T et al (2006) Imaging steps of lymphatic metastasis reveals that vascular endothelial growth factor-C increases metastasis by increasing delivery of cancer cells to lymph nodes: therapeutic implications. Cancer Res 66:8065–8075
Hwang-Bo J, Hyun Yoo K, Park JH, Jeong HS, Chung IS (2012) Recombinant canstatin inhibits angiopoietin-1-induced angiogenesis and lymphangiogenesis. Int J Cancer 131:298–309
Isaka N, Padera TP, Hagendoorn J, Fukumura D, Jain RK (2004) Peritumor lymphatics induced by vascular endothelial growth factor-C exhibit abnormal function. Cancer Res 64:4400–4404

Issa A, Le TX, Shoushtari AN, Shields JD, Swartz MA (2009) Vascular endothelial growth factor-C and C-C chemokine receptor 7 in tumor cell-lymphatic cross-talk promote invasive phenotype. Cancer Res 69:349–357

Jain RK, Fenton BT (2002) Intratumoral lymphatic vessels: a case of mistaken identity or malfunction? J Natl Cancer Inst 94:417–421

Jakob C et al (2011) Lymphangiogenesis in regional lymph nodes is an independent prognostic marker in rectal cancer patients after neoadjuvant treatment. PLoS One 6:e27402

Jeon BH et al (2008) Profound but dysfunctional lymphangiogenesis via vascular endothelial growth factor ligands from CD11b+ macrophages in advanced ovarian cancer. Cancer Res 68:1100–1109

Juarranz A, Jaen P, Sanz-Rodriguez F, Cuevas J, Gonzalez S (2008) Photodynamic therapy of cancer. Basic principles and applications. Clin Transl Oncol 10:148–154

Karlsen TV, McCormack E, Mujic M, Tenstad O, Wiig H (2012) Minimally invasive quantification of lymph flow in mice and rats by imaging depot clearance of near-infrared albumin. Am J Physiol Heart Circ Physiol 302:H391–H401

Karnezis T et al (2012) VEGF-D promotes tumor metastasis by regulating prostaglandins produced by the collecting lymphatic endothelium. Cancer Cell 21:181–195

Kodera Y et al (2011) Sunitinib inhibits lymphatic endothelial cell functions and lymph node metastasis in a breast cancer model through inhibition of vascular endothelial growth factor receptor 3. Breast Cancer Res 13:R66

Kurahara H et al (2010) Significance of lymphangiogenesis in primary tumor and draining lymph nodes during lymphatic metastasis of pancreatic head cancer. J Surg Oncol 102:809–815

Kwon S, Sevick-Muraca EM (2007) Noninvasive quantitative imaging of lymph function in mice. Lymphat Res Biol 5:219–231

Lai G, Rockall AG (2010) Lymph node imaging in gynecologic malignancy. Semin Ultrasound CT MR 31:363–376

Leak LV et al (2004) Proteomic analysis of lymph. Proteomics 4:753–765

Lee JW et al (2007) Peripheral antigen display by lymph node stroma promotes T cell tolerance to intestinal self. Nat Immunol 8:181–190

Leu AJ, Berk DA, Yuan F, Jain RK (1994a) Flow velocity in the superficial lymphatic network of the mouse tail. Am J Physiol 267:H1507–H1513

Leu AJ et al (1994b) Microvascular dynamics in normal skin versus skin overlying Kaposi's sarcoma. Microvasc Res 47:140–144

Leu AJ, Berk DA, Lymboussaki A, Alitalo K, Jain RK (2000) Absence of functional lymphatics within a murine sarcoma: a molecular and functional evaluation. Cancer Res 60:4324–4327

Liao S et al (2011) Impaired lymphatic contraction associated with immunosuppression. Proc Natl Acad Sci USA 108:18784–18789

Liotta LA, Kleinerman J, Saidel GM (1974) Quantitative relationships of intravascular tumor cells, tumor vessels, and pulmonary metastases following tumor implantation. Cancer Res 34:997–1004

Liu J et al (2011) PDGF-D improves drug delivery and efficacy via vascular normalization, but promotes lymphatic metastasis by activating CXCR4 in breast cancer. Clin Cancer Res 17:3638–3648

Lukacs-Kornek V et al (2011) Regulated release of nitric oxide by nonhematopoietic stroma controls expansion of the activated T cell pool in lymph nodes. Nat Immunol 12:1096–1104

Lund A, Duares FV, Hirosue S, Raghavan VR, Nembrini C, Thomas SN, Issa A, Hugues S, Swartz MA (2012) VEGF-C promotes immune tolerance in B16 melanomas and cross-presentation of tumor antigen by lymph node lymphatics. Cell Rep 1:191–199

Lunt SJ et al (2008) Interstitial fluid pressure, vascularity and metastasis in ectopic, orthotopic and spontaneous tumours. BMC Cancer 8:2

Mandriota SJ et al (2001) Vascular endothelial growth factor-C-mediated lymphangiogenesis promotes tumour metastasis. EMBO J 20:672–682

Mannello F, Medda V, Tonti GA (2009) Protein profile analysis of the breast microenvironment to differentiate healthy women from breast cancer patients. Expert Rev Proteomics 6:43–60
Martinez-Corral I et al (2012) In vivo imaging of lymphatic vessels in development, wound healing, inflammation, and tumor metastasis. Proc Natl Acad Sci USA 109:6223–6228
Matsui J et al (2008) Multi-kinase inhibitor E7080 suppresses lymph node and lung metastases of human mammary breast tumor MDA-MB-231 via inhibition of vascular endothelial growth factor-receptor (VEGF-R) 2 and VEGF-R3 kinase. Clin Cancer Res 14:5459–5465
Mattila MM et al (2002) VEGF-C induced lymphangiogenesis is associated with lymph node metastasis in orthotopic MCF-7 tumors. Int J Cancer 98:946–951
Maus EA et al (2012) Near-infrared fluorescence imaging of lymphatics in head and neck lymphedema. Head Neck 34:448–453
Mazzone M et al (2009) Heterozygous deficiency of PHD2 restores tumor oxygenation and inhibits metastasis via endothelial normalization. Cell 136:839–851
McDonald DM, Baluk P (2002) Significance of blood vessel leakiness in cancer. Cancer Res 62:5381–5385
Miteva DO et al (2010) Transmural flow modulates cell and fluid transport functions of lymphatic endothelium. Circ Res 106:920–931
Mittal A et al (2009) The proteome of mesenteric lymph during acute pancreatitis and implications for treatment. JOP 10:130–142
Modi S, Stanton AW, Mortimer PS, Levick JR (2007a) Clinical assessment of human lymph flow using removal rate constants of interstitial macromolecules: a critical review of lymphoscintigraphy. Lymphat Res Biol 5:183–202
Modi S et al (2007b) Human lymphatic pumping measured in healthy and lymphoedematous arms by lymphatic congestion lymphoscintigraphy. J Physiol 583:271–285
Mouli SK, Zhao LC, Omary RA, Thaxton CS (2010) Lymphotropic nanoparticle enhanced MRI for the staging of genitourinary tumors. Nat Rev Urol 7:84–93
Moussai D et al (2011) The human cutaneous squamous cell carcinoma microenvironment is characterized by increased lymphatic density and enhanced expression of macrophage-derived VEGF-C. J Invest Dermatol 131:229–236
Mumprecht V et al (2010) In vivo imaging of inflammation- and tumor-induced lymph node lymphangiogenesis by immuno-positron emission tomography. Cancer Res 70:8842–8851
Nakamura Y et al (2005) Lymph vessel density correlates with nodal status, VEGF-C expression, and prognosis in breast cancer. Breast Cancer Res Treat 91:125–132
Nathan SS et al (2009) Tumour interstitial fluid pressure may regulate angiogenic factors in osteosarcoma. Ann Acad Med Singapore 38:1041–1047
Ng CP, Swartz MA (2003) Fibroblast alignment under interstitial fluid flow using a novel 3-D tissue culture model. Am J Physiol Heart Circ Physiol 284:H1771–H1777
Ng CP, Swartz MA (2006) Mechanisms of interstitial flow-induced remodeling of fibroblast-collagen cultures. Ann Biomed Eng 34:446–454
Ng CP, Helm CL, Swartz MA (2004) Interstitial flow differentially stimulates blood and lymphatic endothelial cell morphogenesis in vitro. Microvasc Res 68:258–264
Ng CP, Hinz B, Swartz MA (2005) Interstitial fluid flow induces myofibroblast differentiation and collagen alignment in vitro. J Cell Sci 118:4731–4739
Niedbala W et al (2007) Nitric oxide induces CD4+CD25+ Foxp3 regulatory T cells from CD4+CD25 T cells via p53, IL-2, and OX40. Proc Natl Acad Sci USA 104:15478–15483
Padera TP et al (2002) Lymphatic metastasis in the absence of functional intratumor lymphatics. Science 296:1883–1886
Patel DD et al (2001) Chemokines have diverse abilities to form solid phase gradients. Clin Immunol 99:43–52
Patel V et al (2011) Decreased lymphangiogenesis and lymph node metastasis by mTOR inhibition in head and neck cancer. Cancer Res 71:7103–7112
Pathak AP et al (2005) Characterizing extravascular fluid transport of macromolecules in the tumor interstitium by magnetic resonance imaging. Cancer Res 65:1425–1432

Peinado H, Lavotshkin S, Lyden D (2011) The secreted factors responsible for pre-metastatic niche formation: old sayings and new thoughts. Semin Cancer Biol 21:139–146

Polacheck WJ, Charest JL, Kamm RD (2011) Interstitial flow influences direction of tumor cell migration through competing mechanisms. Proc Natl Acad Sci USA 108:11115–11120

Psaila B, Lyden D (2009) The metastatic niche: adapting the foreign soil. Nat Rev Cancer 9:285–293

Rasmussen JC, Tan IC, Marshall MV, Fife CE, Sevick-Muraca EM (2009) Lymphatic imaging in humans with near-infrared fluorescence. Curr Opin Biotechnol 20:74–82

Rinderknecht M, Villa A, Ballmer-Hofer K, Neri D, Detmar M (2010) Phage-derived fully human monoclonal antibody fragments to human vascular endothelial growth factor-C block its interaction with VEGF receptor-2 and 3. PLoS One 5:e11941

Roberts N et al (2006) Inhibition of VEGFR-3 activation with the antagonistic antibody more potently suppresses lymph node and distant metastases than inactivation of VEGFR-2. Cancer Res 66:2650–2657

Roozendaal R, Mebius RE, Kraal G (2008) The conduit system of the lymph node. Int Immunol 20:1483–1487

Roozendaal R et al (2009) Conduits mediate transport of low-molecular-weight antigen to lymph node follicles. Immunity 30:264–276

Ruddell A et al (2008) Dynamic contrast-enhanced magnetic resonance imaging of tumor-induced lymph flow. Neoplasia 10:706–713, 701 p following 713

Schomber T et al (2009) Differential effects of the vascular endothelial growth factor receptor inhibitor PTK787/ZK222584 on tumor angiogenesis and tumor lymphangiogenesis. Mol Cancer Ther 8:55–63

Schoppmann SF et al (2006) VEGF-C expressing tumor-associated macrophages in lymph node positive breast cancer: impact on lymphangiogenesis and survival. Surgery 139:839–846

Sevick-Muraca EM (2012) Translation of near-infrared fluorescence imaging technologies: emerging clinical applications. Annu Rev Med 63:217–231

Sevick-Muraca EM et al (2008) Imaging of lymph flow in breast cancer patients after microdose administration of a near-infrared fluorophore: feasibility study. Radiology 246:734–741

Shieh AC, Rozansky HA, Hinz B, Swartz MA (2011) Tumor cell invasion is promoted by interstitial flow-induced matrix priming by stromal fibroblasts. Cancer Res 71:790–800

Shields JD et al (2004) Lymphatic density and metastatic spread in human malignant melanoma. Br J Cancer 90:693–700

Shields J et al (2007a) Autologous chemotaxis as a mechanism of tumor cell homing to lymphatics via interstitial flow and autocrine CCR7 signaling. Cancer Cell 11:526–538

Shields JD et al (2007b) Chemokine-mediated migration of melanoma cells towards lymphatics—a mechanism contributing to metastasis. Oncogene 26:2997–3005

Shields JD et al (2010) Induction of lymphoid-like stroma and immune escape by tumors that express the chemokine CCL21. Science 328:749–752

Sixt M et al (2005) The conduit system transports soluble antigens from the afferent lymph to resident dendritic cells in the T cell area of the lymph node. Immunity 22:19–29

Skobe M et al (2001) Induction of tumor lymphangiogenesis by VEGF-C promotes breast cancer metastasis. Nat Med 7:192–198

Stacker SA et al (2001) VEGF-D promotes the metastatic spread of tumor cells via the lymphatics. Nat Med 7:186–191

Stanton AW et al (2006) A quantitative lymphoscintigraphic evaluation of lymphatic function in the swollen hands of women with lymphoedema following breast cancer treatment. Clin Sci (Lond) 110:553–561

Stohrer M, Boucher Y, Stangassinger M, Jain RK (2000) Oncotic pressure in solid tumors is elevated. Cancer Res 60:4251–4255

Swartz MA, Berk DA, Jain RK (1996) Transport in lymphatic capillaries. I. Macroscopic measurements using residence time distribution theory. Am J Physiol 270:H324–H329

Szuba A, Pyszel A, Jedrzejuk D, Janczak D, Andrzejak R (2007) Presence of functional axillary lymph nodes and lymph drainage within arms in women with and without breast cancer-related lymphedema. Lymphology 40:81–86
Tammela T et al (2011) Photodynamic ablation of lymphatic vessels and intralymphatic cancer cells prevents metastasis. Sci Transl Med 3:69ra11
Teng PN et al (2010) Assessment of buffer systems for harvesting proteins from tissue interstitial fluid for proteomic analysis. J Proteome Res 9:4161–4169
Tong RT et al (2004) Vascular normalization by vascular endothelial growth factor receptor 2 blockade induces a pressure gradient across the vasculature and improves drug penetration in tumors. Cancer Res 64:3731–3736
Truman LA et al (2012) ProxTom lymphatic vessel reporter mice reveal Prox1 expression in the adrenal medulla, megakaryocytes, and platelets. Am J Pathol 180:1715–1725
Tvorogov D et al (2010) Effective suppression of vascular network formation by combination of antibodies blocking VEGFR ligand binding and receptor dimerization. Cancer Cell 18:630–640
Vakoc BJ et al (2009) Three-dimensional microscopy of the tumor microenvironment in vivo using optical frequency domain imaging. Nat Med 15:1219–1223
Wiig H, Tveit E, Hultborn R, Reed RK, Weiss L (1982) Interstitial fluid pressure in DMBA-induced rat mammary tumours. Scand J Clin Lab Invest 42:159–164
Wiig H, Aukland K, Tenstad O (2003) Isolation of interstitial fluid from rat mammary tumors by a centrifugation method. Am J Physiol Heart Circ Physiol 284:H416–H424
Wiig H, Tenstad O, Iversen PO, Kalluri R, Bjerkvig R (2010) Interstitial fluid: the overlooked component of the tumor microenvironment? Fibrogenesis Tissue Repair 3:12
Willett CG et al (2004) Direct evidence that the VEGF-specific antibody bevacizumab has antivascular effects in human rectal cancer. Nat Med 10:145–147
Winkler F et al (2004) Kinetics of vascular normalization by VEGFR2 blockade governs brain tumor response to radiation: role of oxygenation, angiopoietin-1, and matrix metalloproteinases. Cancer Cell 6:553–563
Yang H et al (2011) Soluble vascular endothelial growth factor receptor-3 suppresses lymphangiogenesis and lymphatic metastasis in bladder cancer. Mol Cancer 10:36
Yuan F et al (1996) Time-dependent vascular regression and permeability changes in established human tumor xenografts induced by an anti-vascular endothelial growth factor/vascular permeability factor antibody. Proc Natl Acad Sci USA 93:14765–14770
Zhao J et al (2010) Towards characteristics of photodynamic drugs specifically aimed at microvascular diseases. Mini Rev Med Chem 10:332–341
Zilberberg L et al (2012) Specificity of latent TGF-β binding protein (LTBP) incorporation into matrix: role of fibrillins and fibronectin. J Cell Physiol 227:3828–3836

Chapter 8
The Brain and the Lymphatic System

Mark E. Wagshul and Miles Johnston

Abstract The brain lacks a local lymphatic system, primarily due to the closed environment of the skull which sets strict requirements for control of fluid balance and intracranial pressure. Proper fluid and pressure balance are maintained in the brain through the unique systems of cerebrospinal and interstitial fluid as well as a tight coupling between these systems and the surrounding lymphatic drainage pathways, primarily in the cervical lymph nodes. In this chapter, we will review the physiology of cerebrospinal and interstitial fluid, provide an overview of their primary production and drainage mechanisms, and discuss the still-debated issue of the interconnections of these systems and their relevance to human physiology. We present the current evidence pointing to the importance of the extracranial lymphatic system as one of the key drainage pathways for cerebrospinal fluid from the brain, and conclude with the implications of these interconnected pathways to the ongoing revision for the concept of immune privilege of the brain.

8.1 Introduction

The most important function of the lymphatic system is to remove extravasated proteins from the tissues, as these cannot be effectively absorbed back into the blood capillaries. Consequently, most of the organs and tissues of the body containing blood vessels also contain lymphatics. The brain is a major exception to this rule,

---

M.E. Wagshul (✉)
Department of Radiology, Gruss MR Research Center, Albert Einstein College of Medicine, Bronx, NY, USA
e-mail: mark.wagshul@einstein.yu.edu

M. Johnston
Department of Laboratory Medicine and Pathobiology, University of Toronto, Sunnybrook Research Institute, Toronto, Canada

L. Santambrogio (ed.), *Immunology of the Lymphatic System*,
DOI 10.1007/978-1-4614-3235-7_8, © Springer Science+Business Media New York 2013

lacking a local lymphatic drainage system. As we will show, there are a number of important reasons for this distinction, mostly relating to the strict volume and pressure maintenance requirements, the tight control maintained by the blood–brain barrier (BBB), as well as the so-called immune privilege of the brain. At the same time, we will also show that the lymphatic system in the head and neck (primarily the cervical lymph nodes) still play an important role in fluid drainage and immune function of the brain.

In most tissues, extracellular fluid is commonly formed by filtration. Hydrostatic capillary pressure favors filtration, while colloid osmotic pressure from the plasma proteins oppose filtration. In most tissues, the forces across the capillary favor filtration, forcing the interstitial fluid (ISF) along with its protein and other solutes to be removed by the lymphatic system. In the brain, however, the capillary endothelial cells and their intercellular junctions form a tight barrier to proteins; filtration from the vasculature is very low and the extracellular fluid has relatively little protein compared to other organs. This creates a situation which is the reverse of the normal capillary wall: a very large effective osmotic pressure creating the dominant driving force, with hydrostatic pressure being relatively unimportant, across the normal BBB. In resting states, the BBB is the most important factor regulating the volume of the brain and spinal cord. The low hydraulic conductivity of central nervous system (CNS) capillaries and the high osmotic activity of the major solutes in the blood–brain system provide tight control of CNS volume (Fenstermacher 1984; Bradbury 1984, 1985). The more important implication for our purposes is that the low protein tissue content means that there is no true lymph within the CNS tissue, and thus no need for a true lymphatic system. In addition to these passive transport mechanisms, endothelial carrier transport systems exist to supply the brain with compounds required to maintain metabolic and electrophysiological activities.

The brain is also unique in that in addition to the ISF it has a second, interconnected extracellular fluid compartment, the cerebrospinal fluid (CSF), which plays a critical role in fluid balance and transport. The CSF also serves the role of providing buoyancy of the brain within the skull for protection. While this is an important function, most anatomy textbooks unfortunately stress this function over fluid maintenance and transport which is likely a more critical function of the CSF within the unique intracranial environment. As opposed to the ISF which permeates the intercellular spaces of the white and gray matter, CSF is primarily contained within two distinct fluid compartments in the brain, the ventricles and the subarachnoid spaces surrounding the brain. Ultimately, the parenchymal interstitium connects with these CSF compartments through the ependymal and pia-glial membranes, and to a lesser extent through the Virchow-Robin, or perivascular, spaces (Fenstermacher 1984). Estimates of exchange between ISF and CSF are difficult to make, but are approximately 10 % (ISF leakage into the CSF spaces) in the rat (Szentistvanyi et al. 1984). The two fluid environments thus appear to operate as two separate and distinct entities, with separate and distinct production, drainage and functional mechanisms (Weller 1998). So, as a first look, these two environments will be described separately with the goal of understanding the origins of each fluid, their main

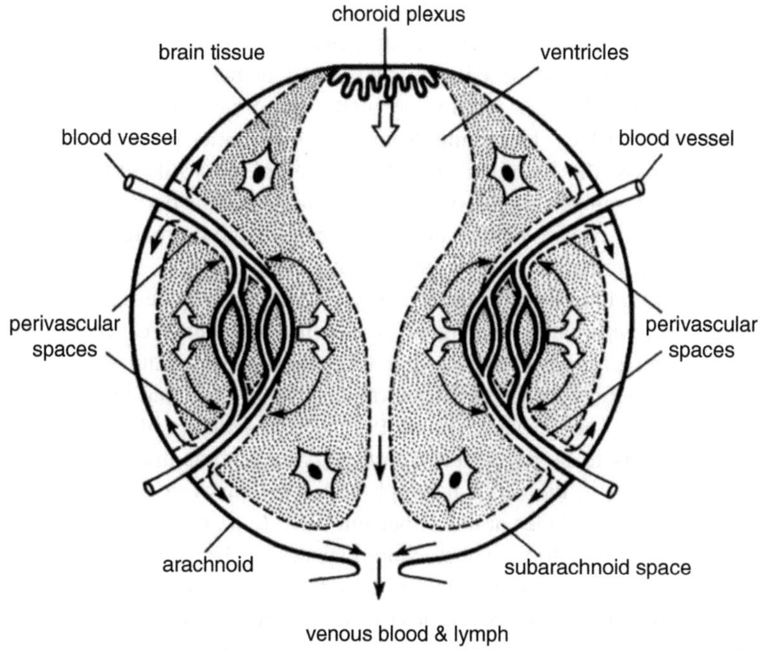

Fig. 8.1 Schematic of the fluid pathways in the brain. The two, mostly independent, fluid pools within the brain are interstitial fluid (ISF) and cerebrospinal fluid (CSF). ISF is produced within the brain, primarily at the capillary endothelium, and drains along perivascular spaces either directly into the cervical lymphatics or into the CSF-filled ventricles and subarachnoid spaces. CSF is produced by the choroid plexus within the cerebral ventricles and drains through the subarachnoid spaces either into venous system via the arachnoid villi and granulations or through the cribriform plate and nasal submucosa into the cervical lymphatics. Reprinted with permission from (Cserr et al. 1992a)

functions with respect to fluid balance and protein/solute transport, and their drainage pathways. Most important will be the relationship of each fluid drainage pathway to the cervical lymphatic system, and implications of brain-lymphatic pathways with respect to the immunological privilege of the brain. Figure 8.1 shows the anatomical relationships of these two fluid compartments, along with the proposed fluid drainage pathways.

8.2 Interstitial Fluid

There are three likely sites for production of ISF: the capillary endothelium, parenchymal metabolism, and recycled CSF, with the former being most likely main source under normal conditions (Abbott 2004). Bulk ISF flow is either via perivascular spaces (Weller 1998; Abbott 2004; Rennels et al. 1985, 1990) or along white matter tracts (Abbott 2004; Geer and Grossman 1997); because of the tight control of the extracellular space in gray matter and the narrow spaces between cells in the neuropil, bulk ISF flow within gray matter is more restricted and less effected by edema (Weller 1998). ISF drains from the white matter primarily along white matter tracts, across the ependyma and into the ventricles, while gray matter drainage is primarily along periarterial pathways, draining into the perivascular spaces of the leptomeningeal arteries on the surface of the brain and ultimately into the cervical lymphatics (Szentistvanyi et al. 1984; Zhang et al. 1992).

The association between ISF and the cervical lymphatic system has been known for a long time, with early investigations showing an extensive network of "prelymphatic" collecting vessels surrounding the cerebral arteries and ultimately draining within perivascular spaces of the internal carotid arteries and jugular veins into the true lymphatics of the neck (Foldi 1975, 1996; Foldi et al. 1966). As noted above, this appears to be the primary outflow path, with as much as 50 % of the tracer accumulating in the cervical lymph nodes (Szentistvanyi et al. 1984; Cserr et al. 1992a). The remainder drains either into CSF (e.g., transependymal or along perivascular channels into the subarachnoid spaces) or by vascular routes into the venous system (i.e., a route similar to CSF absorption, (Oi and Di Rocco 2006)). The exact location at which brain ISF comes in contact with the lymphatic system is still unknown, but the presence of tracer in the walls of intracranial arteries and their absence from the carotid artery in the neck suggest that the contact occurs at the base of the skull (Szentistvanyi et al. 1984; Weller et al. 2009a). Until it is determined otherwise, it seems likely that a portion of the brain ISF convects into extracranial tissue spaces first and then enters proximal lymphatics, unlike CSF which has direct connection with the nasal lymphatics. Whether or not this situation is in any way similar to ISF drainage in humans is still debated (Weller et al. 2009a, b, 2010).

Of course, because this fluid exists within more confined spaces compared to the large CSF-containing cisterns and ventricles of the brain, demonstration of bulk ISF flow has been more challenging and difficult to quantify. Perivascular outflow channels were demonstrated over 100 years ago (His 1865; Weed 1914). Radioactive tracer injected into brain parenchyma appears abundantly within the perivascular spaces; quantification studies have demonstrated typical ISF flow rates of 0.1–0.3 μL/g/min in rat brain (Szentistvanyi et al. 1984; Cserr and Ostrach 1974; Abbott et al. 1985), or about a factor of 10 less than typical CSF flow rates in the same species (Harnish and Samuel 1988). Interestingly, this ISF flow rate is comparable to the ISF drainage rate in the rest of the body, where ISF drains directly into the lymphatic system. More recent techniques utilizing multiple labeled immunofluorescent imaging have shown co-localization of the tracer with the laminin within capillary and arterial basement membrane (Weller et al. 2009a), prompting the labeling of vessel basement membranes as the "lymphatics of the brain" (Carare et al. 2008).

The source of the motive force driving perivascular ISF flow is still debated. It has been suggested that flow is driven by arterial pulsations forces, into the brain along periarterial spaces and out toward the brain surface along perivenule spaces

(Rennels et al. 1985, 1990). ISF drainage was significantly diminished in these studies by aortic occlusion or partial ligation of the brachiocephalic artery. More recently, investigators have suggested that the deposition patterns typically seen in Alzheimer's disease patients can be explained by ISF drainage along a periarterial pathway (Weller 1998; Weller et al. 2008; Preston et al. 2003), providing detailed mathematical modeling to show that the pulsatile arterial pumping action can produce drainage of solutes (in the direction counter to the blood flow) along the perivascular space (Schley et al. 2006). An intriguing element of this theory, although as yet completely speculative, is that amyloid-beta accumulation with aging and the significant associations with atherosclerotic disease (Dolan et al. 2010), might then be explained by the reduced pulsatile pumping action of the hardened arteries in this population.

8.3 Cerebrospinal Fluid

CSF is a clear fluid which occupies the ventricles and subarachnoid spaces within the brain and surrounding the spinal canal. It is produced in the human at a rate of about 500 mL/day, by passive filtration and active transport from the blood within the choroid plexi of the lateral, third and fourth ventricles; other sources are blood vessels and the ependymal walls of the lateral ventricles (Milhorat 1987; Hammock and Milhorat 1976). Regulation of the CSF production rate is critical for maintaining brain fluid homeostasis, plays an important role in determination of intracranial pressure (ICP) and changes in production rate have been reported with development of hydrocephalus and with aging (Johanson et al. 2008; Silverberg et al. 2002). CSF production has been compared by some to a renal-like function, providing fluid stabilization for the CNS (Johanson et al. 2008; Spector and Johanson 1989). From these production sites, the CSF flows through the ventricles, into the subarachnoid spaces via the foramina of Luschka and of Magendie and circulates over the outer surfaces of the brain and spinal cord. Studies using the ventriculocisternal or ventriculolumbar perfusion methods have demonstrated the dependence of CSF absorption on CSF pressure (Rubin et al. 1966; Cutler et al. 1968; Bering and Sato 1963; Heisey et al. 1962).

8.3.1 Pulsatile CSF Flow

While bulk flow of CSF through this system is important for maintaining normal pressure, flow and CSF turnover, there is another component to the flow which is unique to intracranial fluid dynamics. Approximately once a second, shortly following cardiac systole, a pulse of arterial blood enters the cranium. While pulsatile arterial blood flow exists everywhere in the body, this pulse is quickly dissipated as it is transmitted down the vascular tree. By the time the pulse reaches the capillary bed, it has been completely dissipated, resulting in smooth capillary and venous flow. The dissipation is facilitated by the transfer of the pulse wave into the surrounding compliant tissue. Because of the unique mechanical environment of the cranium, which is (almost completely) enclosed within the rigid skull, dissipation of the arterial pulse wave is more complex. In this case, the pulse wave is transferred into the fluids surrounding the incoming arteries, i.e., the CSF within the subarachnoid spaces. Part of the pulse wave reaches the deeper tissue and is transferred into the brain tissue, ultimately being transmitted into the CSF-filled ventricles and subarachnoid spaces. In turn, the CSF dissipates the pulse wave by transmitting it either into the venous system, or into the compliant subarachnoid spaces of the spine, where the pulse waves are transmitted into the surrounding tissues (Wagshul et al. 2011).

This unique environment produces a significant degree of pulsatile CSF flow within the ventricles and subarachnoid spaces. MRI can been used to quantify these flows, and has shown that pulsatile CSF flow exceeds the bulk flow described above by an order of magnitude or more (Nitz et al. 1992; Naidich et al. 1993; Bhadelia et al. 1995). Measurement of pulsatile CSF flow is often used in hydrocephalus as a marker of disease severity and as a prognostic for predicting outcome from shunt surgery (Poca et al. 2002; Bradley et al. 1996; Baledent et al. 2004; Luetmer et al. 2002). While it has not been determined if pulsatile CSF flow affects overall CSF turnover and drainage, the pulsatile motion of CSF within the brain has the effect of enhancing fluid mixing within the CSF spaces (Greitz 1993; Bulat and Klarica 2011; Vladic et al. 2009). These effects likely play an important role in determining CSF distributions, as the rapid transport of tracer out of the CSF compartments are difficult to explain based on bulk drainage rates alone (Rennels et al. 1985, 1990).

8.3.2 The Arachnoid Projections

With respect to the absorption site, the classical view is that bulk CSF drainage occurs largely through the arachnoid projections (Glimcher et al. 2008; Pollay 2010; Welch and Pollay 1961). The arachnoid villi and granulations are herniations of the arachnoid membrane that penetrate gaps in the dura and protrude into the venous sinuses. The villi are only seen microscopically, but the granulations are large, macroscopic structures (reviewed in (Pollay 2010; Davson et al. 1987)). Arachnoid villi are not restricted to the intracranial venous sinuses. There are clusters of arachnoidal cells adjacent to the emerging spinal nerve roots that are attached to and also penetrate the dura of the root sleeves extending into small spinal veins (Gomez et al. 1974).

Quantitative support for these structures comes mainly from in vitro and ex vivo studies involving the positioning of dura mater-containing arachnoid projections or cells from arachnoid granulations grown on a filter as occluding membranes between two chambers. These studies demonstrate transport from the subarachnoid side to the venous side (or basal to apical direction in the case of cells) but little or no transport in the opposite direction (Glimcher et al. 2008; Welch and Pollay 1961; Welch and Friedman 1960; Pollay and Welch 1962; Grzybowski et al. 2006). However, there is surprisingly little in vivo evidence to support a role for the arachnoid projections in normal CSF absorption and their function is now being questioned (Egnor et al. 2002; Zakharov et al. 2004a, b; Johnston and Papaiconomou 2002; Papaiconomou et al. 2002). In fact, the most relevant in vivo quantitative data in cats, rabbits, monkeys would seem to suggest that CSF transport into the cranial venous system does *not* occur, at least at normal ICPs (Zakharov et al. 2004a; McComb et al. 1982, 1984; Zlokovic et al. 1990; Papaiconomou et al. 2004), and the arachnoid projections may only be recruited as CSF absorption sites when ICP is elevated.

8.3.3 Aquaporins

For those of us trained in the classical concepts, the majority of extracellular water is removed from the brain by bulk flow; viewed by many as a function of the arachnoid projections (although increasingly considered a task for the lymphatic circulation, see below). Recent research into the importance of aquaporin-4 (AQP-4) in the regulation of cerebral water balance, by active transport across the cell membrane as opposed to bulk transport, would also implicate these sites as important in brain fluid absorption (Owler et al. 2010; MacAulay and Zeuthen 2010; Mao et al. 2006; McAllister and Miller 2006). AQP-4 is the predominant water channel in the brain (Badaut et al. 2002) and is expressed on endothelial cells and astrocytic foot processes that surround capillary endothelial cells (Kobayashi et al. 2001; Bloch and Manley 2007). These proteins appear to function as bidirectional channels to facilitate water movement, driven by hydrostatic or osmotic forces (Bloch and Manley 2007; Nagelhus et al. 2004).

However, it would appear from analysis of AQP-4 null mice that these channels do not serve as the default absorption mechanism for brain water. These experiments revealed that the loss of this gene is well tolerated with no overt defects during development (Manley et al. 2000). The importance of AQP-4 in water regulation only becomes apparent when null animals are challenged with various stresses. AQP-4 null mice display reduced cellular swelling in models of cytotoxic cerebral edema and increased brain swelling in models of vasogenic edema, strongly indicating that AQP-4 plays a role in parenchymal water homeostasis in the brain, at least under pathophysiological conditions (Papaiconomou et al. 2004; Bloch and Manley 2007; Verkman et al. 2006). The question of which forces dominate aquaporin transport has not been worked out, but it seems unlikely that hydrostatic pressure gradients can regulate capillary network fluid uptake since capillary pressures are higher than those in the brain interstitium. Even if the tissue pressures were greater than that in the capillaries, interstitial solutes would be held up at the capillary membrane by the BBB. This would increase the concentration of solutes at this location and increase the pericapillary osmotic pressure, which would limit further water absorption; a process termed osmotic buffering (Fenstermacher 1984). Osmotic forces on the other hand, may be the key to this water absorption although how this might occur is speculative.

8.3.4 Lymphatic Drainage of CSF

The absence of lymphatic vessels within the brain and spinal cord parenchyma would logically seem to exclude this system from any role in CSF absorption. However, CSF-lymphatic connections have been appreciated for over 100 years (reviewed in (Koh et al. 2005)). In general terms, studies in the late nineteenth and early twentieth century demonstrated that various dyes and tracers injected into the CSF or brain parenchyma made their way into lymphatic vessels external to the cranium and into a variety of lymph nodes in the head and neck. One of the most important interfaces between the CSF and lymph compartments is centered at the level of the cribriform plate. CSF convects through the foramina of the cribriform plate along the subarachnoid space surrounding the olfactory nerves and is absorbed by lymphatic vessels that are associated with the nerve roots in the olfactory submucosa. More recently, qualitative and quantitative analyses have revealed a significant role for extracranial lymphatic vessels in CSF drainage (Zakharov et al. 2004a, b; Papaiconomou et al. 2004; Koh et al. 2005, 2006; Nagra et al. 2006; Johnston et al. 2004; Johnston 2003).

8.3.5 Anatomical Brain-Lymphatic Relationships

In most organs and tissues outside the brain, lymphatics take up water and solutes from the interstitium, i.e., the interstitial space is an intervening compartment between the microvasculature, the major source of the material that will become lymph, and the lymphatic vessels. In contrast, CSF and extracranial lymph seem to be in fluid continuity with one another. In rats, for example, CSF moves directly from the subarachnoid space into submucosal lymphatics at the level of the cribriform plate (Zhang et al. 1992; Kida et al. 1993, 1994, 1995). In addition, Microfil infused into the subarachnoid compartment (in mice, rats, rabbits, sheep, pigs, monkeys, and ex vivo in humans) enters an extensive lymphatic network adjacent to the extracranial surface of the cribriform plate (Zakharov et al. 2003, 2004b; Johnston et al. 2004). Lymphatics filled with Microfil were especially conspicuous around the olfactory nerves close to the point of exit from the cribriform plate.

Multiple lymphatic ducts form a "collar" around the olfactory nerve providing a barrier that separates the perineurial space (and CSF convecting from the subarachnoid compartment) from the surrounding submucosal tissues at the emerging nerve rootlets. This "direct" connection between lymphatic endothelial cells and the outer cell layer of the olfactory nerves would be expected to expedite CSF removal from

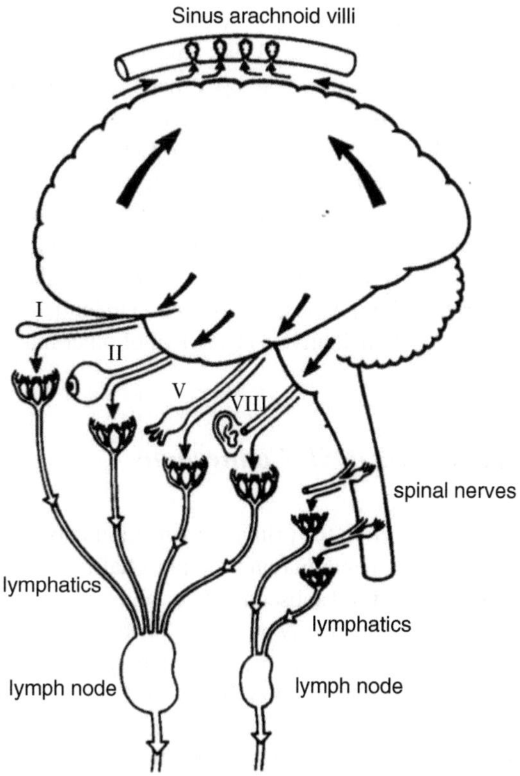

Fig. 8.2 Outflow pathways from the cranial and spinal subarachnoid spaces, showing the important lymphatics connections. Although CSF is traditionally through to flow out exclusively through the arachnoid villi into the venous sinuses, evidence for the importance of the lymphatic drainage pathways has been accumulating over the last 30 years, with outflow along cranial nerves and spinal nerve roots (I—olfactory, II—optic, V—trigeminal, VIII—acoustic). Reprinted with permission from (Cserr et al. 1992a)

the subarachnoid space. Fluid continuity between the CSF and extracranial lymph compartments would allow CSF transport under a wide variety of ICPs and presumably, help to protect the brain from air-borne infection by limiting CSF access to well-defined lymphatic drainage pathways rather than permit random CSF dispersion throughout the extracellular spaces of the olfactory and respiratory submucosa (in a manner similar to direct peritoneal fluid transport into diaphragmatic lymphatics without first passing into the diaphragmatic interstitium (Ohtani et al. 2001; Shinohara et al. 2003)). Figure 8.2 shows the CSF outflow pathways, including the major lymphatic connections.

CSF-lymphatic connections do not appear to be specific to any one species, being seen in mice (Johnston et al. 2004; Moinuddin and Tada 2000), rats (Zhang et al. 1992; Nagra et al. 2006; Johnston et al. 2004; Kida et al. 1993, 1994; Boulton et al. 1999; Brinker et al. 1994a, 1997), guinea pigs (Arnold and Ilberg 1973), pigs (Johnston et al. 2004), rabbits (Johnston et al. 2004; Erlich et al. 1986; Bradbury et al. 1981; Yamada et al. 1991; Gomez et al. 1985), cats (McComb et al. 1984; Brinker et al. 1994a), dogs (Brinker et al. 1994a, b), sheep (Zakharov et al. 2003, 2004b; Papaiconomou et al. 2004; Johnston et al. 2004) and nonhuman primates (Johnston et al. 2004, 2005; Brinker et al. 1994b; McComb and Hyman 1990). There is also ex vivo evidence that such CSF-lymphatic connections exist in humans as well (Johnston et al. 2004; Weller et al. 1992; Csanda et al. 1983; Caversaccio et al. 1996; Lowhagen et al. 1994; Rudert and Tillmann 1993). Indian ink administered into the CSF in human autopsy material fills the perineurial spaces around the olfactory nerve branches and was found in the nasal submucosal tissue (Lowhagen et al. 1994). Red blood cells were also observed around the olfactory nerves and within the nasal mucosa post-subarachnoid hemorrhage. Perhaps most persuasively, Microfil administered into the cisterna magna in cadaveric material was observed adjacent to the cribriform plate and extending into lymphatics in the olfactory submucosa and on the nasal septum (Johnston et al. 2004), similar to in vivo observations in animals. Therefore, lymphatic CSF absorption in human and nonhuman primates is not different (at least qualitatively) from that of other mammalian species.

8.3.6 *ISF-CSF Connections*

As noted earlier, the primary connection between the ISF and CSF systems is at the ventricular ependyma and the pia-glial boundary of the subarachnoid space, with some of this fluid subsequently draining into cervical lymphatics through the cribriform plate. This likely accounts for the 10–15 % efflux of radioactive tracer recovered from bulk CSF in the cisternum magnum after injection in caudate and internal capsule (Szentistvanyi et al. 1984). However, some studies indicate that there are also direct connections between parenchymal ISF and extracranial lymph, similar to CSF. In these same experiments, flow of ISF into bulk CSF accounted for 60–75 % of efflux from midbrain. In addition, tracer injected into brain appears primarily in the ipsilateral cervical lymph node, while injection into CSF appears equally on both sides, suggesting that the majority of ISF does not exit the brain via a well-mixed CSF pool (Bradbury et al. 1981). Furthermore, higher cervical lymph tracer concentrations (normalized to CSF) were observed after brain injection than were noted with administration into the CSF. As discussed above, it would appear that a large fraction of the ISF drains through the subarachnoid space via adventitial perivascular channels (Casley-Smith et al. 1978). Similar to the question of lymphatic uptake in humans, the issue of ISF-CSF communication in humans remains an open question, although it is interesting to note that CSF levels of brain proteins, which would require such communication, are commonly used as biomarkers in a number of dementia disorders (Mollenhauer and Trenkwalder 2009; Eller and Williams 2009; Frankfort et al. 2008; Leinonen et al. 2011).

8.3.7 *Quantitative Analysis of Lymphatic CSF Absorption*

While anatomical CSF-lymphatic connections are interesting and likely relevant to actual fluid uptake, quantitative methods are necessary in order to ascertain the relative importance of uptake through the cribriform plate and into the ethmoidal lymphatic system, relative to other absorption pathway (e.g., the arachnoid granulations). Estimating the transport of an intra-cisternally administered radioactive tracer into these ducts would provide such a measure. Similarly, the thoracic duct can provide data on the entry of spinal CSF into the lymphatics. The level of plasma recovery of radiolabeled human serum albumin (^{125}I-HSA), injected into the lateral ventricles, was found to change by 100 % when all relevant lymphatic pathways were interrupted (Boulton et al. 1998, 1999), indicating that lymphatic vessels transport approximately 50 % of the protein tracer from the CSF compartment into plasma. Similar results were obtained from studies in sheep and rats. Further, detailed quantitation correcting for errors introduced by filtration suggest that 40–48 % of all CSF removed from the cranial compartment in sheep is cleared through the cribriform plate and into extracranial lymphatics (Boulton et al. 1998). Therefore, in sheep at least, the extracranial lymphatic vessels downstream of the cribriform plate are responsible for nearly half of global CSF absorption.

There would also appear to be CSF-lymphatic connections within the spinal subarachnoid spaces (about 25 % of CSF drainage in sheep occurs through the spinal subarachnoid space (Bozanovic-Sosic et al. 2001)). Injection of Evans blue dye into the spinal subarachnoid space results in the visualization of peri-spinal lymphatics vessels and stained intercostal lymph nodes (Koh et al. 2005). In addition, CSF outflow resistance is almost threefold higher in the spinal subarachnoid space compared to the cranial compartment, indicating a much higher resistance to absorption from the spinal compartment. This would seem to fit the observation that spinal uptake is indirect, convecting first into the peri-spinal interstitial space before being taken up by regional lymphatic vessels, compared to the direct CSF linkage in the cranium. While arachnoid proliferations resembling the villi and granulations of the cranial system have been described in spinal tissues (Gomez et al. 1974; Welch and Pollay 1963; Kido et al. 1976), their significance is questionable. Arachnoid projections in monkey are associated directly with veins in only 5 of the 32 nerve roots investigated (Welch and Pollay 1963).

Thus, of the net CSF absorption it would appear that (at least in nonhuman vertebrates) only 25 % is unaccounted for, which could possibly relate to the arachnoid projections or other cranial or spinal nerves.

8.3.8 Development of CSF-Lymph Connections

One would expect that CSF outflow pathways develop or decline in parallel with CSF formation rates over time. Paradoxically, this does not appear to be the case with the arachnoid projections. The choroid plexus develops relatively early in gestation, and by the third gestational month it nearly fills both of the lateral ventricles (Ce 1995). At the same time, in two microscopic studies of autopsy specimens from individuals up to 56 days old (Osaka et al. 1980) and from the 18th week of gestation to 80 years (Fox et al. 1996), no arachnoid villi or granulations are observed before birth. Arachnoid projections start to become visible in the dura at or around the time of birth (Osaka et al. 1980). In another study, arachnoid villi were present at the 35th week and granulations by the 39th week (Gomez et al. 1982). These data imply that the neonate must have a mechanism for CSF absorption other than via arachnoid villi. The problem is further confounded by the changes with age: the villi and granulations increase in number with age, and are in abundance in adults. Given the progressive decline in CSF production in the elderly (Preston 2001), it is not clear why greater numbers of arachnoid projections are needed in the aging population.

There is little information on the developmental relationships between the lymphatic circulation and CSF formation. Nonetheless, there is some evidence that the appearance of CSF-lymphatic connections and CSF secretory capacity are related. CSF-lymphatic connections have been investigated using Microfil and a soluble Evan's blue-protein complex in two species, pigs (where significant choroid plexus CSF synthesis begins before birth) and rats (where CSF secretion is markedly up regulated within the first weeks after birth) (Koh et al. 2006). In pigs, CSF-lymphatic connections do not appear to exist at E80-81, but may develop as early as E92, while these associations are not obvious in rats until about 1 week after birth. These data suggest that at least in these two species extracranial lymphatic vessel uptake of CSF develops around the same time that significant volumes of CSF are being produced by the choroid plexus. Additionally, analysis of the uptake of a CSF protein tracer absorption in 3, 6, 12, and 19 months old rats, revealed that lymphatic CSF absorption declines significantly in the older animals mirroring the known drop in CSF formation associated with advancing age (Nagra and Johnston 2007). While not conclusive evidence, the limited available data seems to correlate the lifespan up- and down-regulation in CSF formation with the up- and down-regulation of lymphatic CSF connections.

8.3.9 Relationship Between ICP and Cervical Lymph Flow

One of the important functions of the tight regulation of CSF production and absorption is the maintenance of ICP (Johanson et al. 2008; Johanson 2008), with the most obvious implications in the development of elevated ICP in obstruction hydrocephalus. Given the connection between CSF and extracranial lymphatics, one would expect that elevations in ICP to be reflected by increases in lymph flow from the downstream cervical vessels. In cats, cervical lymph flow rates increase as CSF pressures are raised following infusion of artificial CSF into the cisterna magna (Love and Leslie 1984). In sheep, elevation of ICP by ventriculocisternal perfusion results in significant increases in cervical lymphatic pressures and flow rates (Boulton et al. 1998; Silver et al. 1999). A similar effect is observed in rats, with a dampened response due to loss of protein-free water into the surrounding blood capillaries in the ethmoidal tissues and nodes (Koh et al. 2007). In any event, these studies provide further support for the connection of the CSF and lymphatic systems, clearly showing a hydraulic coupling between CSF and cervical lymph in various species.

The strongest support to this hydraulic coupling comes from a series of experiments in which the CSF absorption pathway through the nasal lymphatics was blocked, resulting in negative consequences for the brain (Mollanji et al. 2001a, 2002). Procedures were developed to seal the cribriform plate on the nasal mucosal side, thereby avoiding complications associated with intracranial access. ICP was approximately doubled post-obstruction (Mollanji et al. 2002), CSF clearance was reduced (Mollanji et al. 2001a) and CSF outflow resistance was elevated (Silver et al. 2002). By reducing CSF absorption from the cranial CSF system, the cranial pressure regulating mechanisms were forced to work further up the pressure-volume curve where compliance was reduced, as was indicated by the increase in amplitude of the ICP pulsations (Marmarou et al. 1975). These data show the direct negative impact of a lymphatic CSF absorption deficit on cranial CSF dynamics and again illustrate the importance of the lymphatic pathway for CSF uptake.

8.3.10 Are the Lymphatics Implicated in the Pathogenesis of Hydrocephalus?

Hydrocephalus is the prototypical CSF disorder. While there is increasing discontent with the classical view that this disease represents a "plumbing problem" with an obstruction to CSF flow or absorption at its core (Bergsneider et al. 2006), it is generally accepted that a CSF drainage deficit is at least a co-conspirator in its pathogenesis. The argument which has been laid out to this point implicates the lymphatic system as an important player in CSF drainage, leading to an obvious question: is there any evidence linking hydrocephalus with disordered lymphatic function? While not definitive, there are several observations that would appear to support the concept that disruption of CSF transit through the cribriform plate may be associated with hydrocephalus. For example, India ink moves more slowly from the CSF into cervical lymph nodes in a mouse model of TGFß1 induced hydrocephalus relative to controls (Moinuddin and Tada 2000), suggesting that the cribriform-lymphatic connections are disrupted. The lack of development of the olfactory bulbs in mice is associated with hydrocephalus (Naruse and Ueta 2002). One might speculate that the olfactory neurons may also be absent or defective in these examples and if this is the case, the important lymphatic connections in the vicinity of the cribriform plate may not exist.

The possible interaction between lymphatic function and hydrocephalus has been explored more directly in a kaolin-induced communicating hydrocephalus model in rats (Nagra et al. 2008). Kaolin administration reduced the lymphatic uptake of a CSF protein tracer significantly. Furthermore, less movement of tracer into the turbinate lymphatics was associated with more severe ventricular dilation. However, it was possible that while lymphatic CSF uptake was compromised, other

CSF absorption pathways may have compensated to the extent that global CSF absorption might be relatively unchanged. In a separate study, CSF outflow resistance was observed to be significantly greater in the kaolin group compared with animals receiving saline suggesting that the lymphatic deficit did indeed have a global impact on CSF absorption (Nagra et al. 2010).

A common objection to these views from the clinical community is that sealing the cribriform plate clinically does not lead to hydrocephalus. However, the physiological and surgical realities are very complex and several factors must be considered. First, 8 % of patients that undergo cranial base surgery for tumors do develop hydrocephalus (Duong et al. 2000) but the mechanism has never been determined. Following repair of CSF leaks (after traumatic brain injury, for example) it is impossible to determine whether all or only a portion of the "olfactory pathways" that link the CSF and lymph compartments have been obliterated. Sealing the cribriform plate may also lead to the recruitment of other lymphatic absorption sites, especially those associated with the spinal subarachnoid compartment (Bozanovic-Sosic et al. 2001; Mollanji et al. 2001b). Indeed, CSF convects along all cranial and spinal nerves (Bradbury and Cserr 1985) and at this point, it is impossible to know how these pathways are affected when the cribriform plate has been blocked. The movement of CSF through the foramina of the cribriform plate into olfactory lymphatic vessels is the paradigm for lymphatic CSF transport and has the practical advantage that absorption at this location can be visualized easily and measured in several ways. It seems likely that any impediments that occur at this location may be mirrored by similar deficits at other sites at which CSF and lymph come in contact.

8.3.11 Immunological Implications of Lymphatic CSF Drainage

The fact that protein tracers injected into the brain and CSF travel to the lymph nodes in the neck would seem to add an immunological perspective to this discussion. Indeed, available evidence would suggest that the brain is not the immunologically privileged site it was once believed to be (Galea et al. 2007; Romo-Gonzalez et al. 2012; Engelhardt and Ransohoff 2005; Engelhardt 2008). The concept of immune privilege certainly does not apply to the CSF, choroid plexus, and meninges. For example, dendritic cells injected into CSF migrate to the cervical lymph nodes (but not when injected into brain parenchyma) (Hatterer et al. 2006), a result which is quite understandable in light of the data presented above of significant CSF drainage into the lymphatic system across the cribriform plate. Given the established link between brain ISF and CSF, it is also not surprising that the afferent arm of the immune system for antigen migration from brain tissue is intact; the CSF and perivascular spaces have been suggested as the major transport paths for antigen out of the CNS (Hatterer et al. 2008). Antigen injected into brain parenchyma drains into the cervical lymph nodes, with a strong antibody response (Cserr et al. 1992a, b;

Harling-Berg et al. 1991). A humoral immune response in mice is generated mainly by the deep cervical lymph nodes after injection of sheep red blood cells into various intracerebral sites (Widner et al. 1988). In rats, infusion of human serum albumin into the cranial CSF leads to antibody production by the cervical lymph nodes, and antibody titres in the peripheral circulation are reduced when cervical lymphatics are obliterated (Harling-Berg et al. 1989).

It has furthermore been shown that a modified efferent arm is also operational. Potential access pathways for lymphocytes into the CNS are across the choroid plexus endothelium, across the pial surface within the subarachnoid perivascular spaces (both indirect routes into the CNS), and direct transport into the parenchyma across the BBB (Ransohoff et al. 2003); while nonactivated lymphocytes cannot cross the intact barrier (Hickey 1991; Wekerle 1993), activated T- lymphocytes can. Nonactivated, recirculating lymphocytes migrate normally from blood into the CSF in concentrations similar to those found in subcutaneous lymph (Kleine and Benes 2006). Lymphocytes infused into the CNS can be found in cervical lymph nodes (Weller et al. 1996). Therefore, lymphocytes found in the brain appear to be part of the recirculating lymphocyte pool and do not require activation to enter the CSF (Seabrook et al. 1998). Finally, the brain-lymph axis would seem to play an important role in the generation of experimental autoimmune encephalomyelitis in rats. Removal of the deep and superficial cervical lymph nodes following induction of autoimmune encephalomyelitis reduces the severity of the pathology significantly (Phillips et al. 1997). Some investigators have speculated that lymphatic drainage of brain antigens could conceivably contribute to the pathogenesis of multiple sclerosis (Weller 1998; Weller et al. 2010). Thus, while the brain does not contain a traditional lymph drainage system, regulated communication between the brain fluids and the lymphatic system of the head and neck defines the unique immune response of the CNS.

8.4 Summary

The brain lacks a local lymphatic system, primarily due to the closed environment of the skull which sets strict requirements for control of fluid balance and ICP. This unique environment is regulated by the tight control on the production and absorption of CSF. Although the drainage of CSF from the brain was for many years assumed to occur primarily thorough the arachnoid villi and granulations, there is increasing evidence that a large percentage of absorption occurs via the lymphatics of the head and neck. In addition, drainage of ISF into the lymphatics has been recognized for many years. Thus, the idea of immune privilege of the brain which was always assumed to imply a strict separation of the healthy brain (with an intact blood brain barrier) is also being revisited. The connection of these two primary brain fluid pools to the lymphatic system thus has immunological implications, allowing for immune surveillance of the intracranial environment and providing for an immune response via delivery of lymphocytes into the brain along these pathways.

References

Abbott NJ (2004) Evidence for bulk flow of brain interstitial fluid: significance for physiology and pathology. Neurochem Int 45(4):545–552

Abbott NJ, Bundgaard M, Cserr HF (1985) Tightness of the blood–brain barrier and evidence for brain interstitial fluid flow in the cuttlefish, Sepia officinalis. J Physiol 368:213–226

Arnold W, Ilberg CV (1973) The connections of the cerebrospinal fluid (CSF) with the lymphatic system of the head and neck. In: Mayall RC, Witte MH (eds) Progress in lymphology IV. Plenum Press, Tucson, AZ, pp 57–58

Badaut J, Lasbennes F, Magistretti PJ, Regli L (2002) Aquaporins in brain: distribution, physiology, and pathophysiology. J Cereb Blood Flow Metab 22(4):367–378

Baledent O, Gondry-Jouet C, Meyer ME et al (2004) Relationship between cerebrospinal fluid and blood dynamics in healthy volunteers and patients with communicating hydrocephalus. Invest Radiol 39(1):45–55

Bergsneider M, Egnor MR, Johnston M et al (2006) What we don't (but should) know about hydrocephalus. J Neurosurg 104(3 suppl):157–159

Bering EA Jr, Sato O (1963) Hydrocephalus: changes in formation and absorption of cerebrospinal fluid within the cerebral ventricles. J Neurosurg 20:1050–1063

Bhadelia RA, Bogdan AR, Wolpert SM (1995) Analysis of cerebrospinal fluid flow waveforms with gated phase-contrast MR velocity measurements. AJNR Am J Neuroradiol 16(2):389–400

Bloch O, Manley GT (2007) The role of aquaporin-4 in cerebral water transport and edema. Neurosurg Focus 22(5):E3

Boulton M, Flessner M, Armstrong D, Hay J, Johnston M (1998) Determination of volumetric cerebrospinal fluid absorption into extracranial lymphatics in sheep. Am J Physiol 274(1 Pt 2): R88–R96

Boulton M, Flessner M, Armstrong D, Mohamed R, Hay J, Johnston M (1999) Contribution of extracranial lymphatics and arachnoid villi to the clearance of a CSF tracer in the rat. Am J Physiol 276(3 Pt 2):R818–R823

Bozanovic-Sosic R, Mollanji R, Johnston MG (2001) Spinal and cranial contributions to total cerebrospinal fluid transport. Am J Physiol Regul Integr Comp Physiol 281(3):R909–R916

Bradbury MW (1984) The structure and function of the blood–brain barrier. Fed Proc 43(2):186–190

Bradbury MW (1985) The blood–brain barrier. Transport across the cerebral endothelium. Circ Res 57(2):213–222

Bradbury MWB, Cserr HF (1985) Drainage of cerebral interstitial fluid and of cerebrospinal fluid into lymphatics. In: Johnston MG (ed) Experimental biology of the lymphatic circulation. Elsevier Science Publishers, Amsterdam, pp 355–394

Bradbury MW, Cserr HF, Westrop RJ (1981) Drainage of cerebral interstitial fluid into deep cervical lymph of the rabbit. Am J Physiol 240(4):F329–F336

Bradley WG Jr, Scalzo D, Queralt J, Nitz WN, Atkinson DJ, Wong P (1996) Normal-pressure hydrocephalus: evaluation with cerebrospinal fluid flow measurements at MR imaging. Radiology 198(2):523–529

Brinker T, Botel C, Samii M (1994a) A species comparing radiological study on the absorption of cerebrospinal fluid into the cervical lymphatic system. In: Nagai H, Kamiya K, Ishii K (eds) Intracranial pressure IX. Springer, Tokyo, pp 559–560

Brinker T, Botel C, Rothkotter HJ, Walter GF, Samii M (1994b) The perineural pathway of cerebrospinal fluid absorption into the cervical lymphatic system. Morphological findings in rats, cats, dogs and monkeys. In: Nagai H, Kamiya K, Ishii K (eds) Intracranial pressure IX. Springer, Tokyo, pp 132–135

Brinker T, Ludemann W, Berens von Rautenfeld D, Samii M (1997) Dynamic properties of lymphatic pathways for the absorption of cerebrospinal fluid. Acta Neuropathol 94(5):493–498

Bulat M, Klarica M (2011) Recent insights into a new hydrodynamics of the cerebrospinal fluid. Brain Res Rev 65(2):99–112

Carare RO, Bernardes-Silva M, Newman TA et al (2008) Solutes, but not cells, drain from the brain parenchyma along basement membranes of capillaries and arteries: significance for cerebral amyloid angiopathy and neuroimmunology. Neuropathol Appl Neurobiol 34(2):131–144

Casley-Smith JR, Clodius L, Foldi-Borcsok E, Gruntzig J, Foldi M (1978) The effects of chronic cervical lymphostasis on regions drained by lymphatics and by prelymphatics. J Pathol 124(1):13–17

Caversaccio M, Peschel O, Arnold W (1996) The drainage of cerebrospinal fluid into the lymphatic system of the neck in humans. ORL J Otorhinolaryngol Relat Spec 58(3):164–166

Ce J (1995) Ventricles and cerebrospinal fluid. In: Conn PM (ed) Neuroscience in medicine. J.B. Lippincott Company, Philadelphia, pp 171–196

Csanda E, Obal F, Obal FJ (1983) Central nervous system and lymphatic system. In: Foldi M, Casley-Smith JR (eds) Lymphangiology. Schattauer, New York, pp 475–508

Cserr HF, Ostrach LH (1974) Bulk flow of interstitial fluid after intracranial injection of blue dextran 2000. Exp Neurol 45(1):50–60

Cserr HF, Harling-Berg CJ, Knopf PM (1992a) Drainage of brain extracellular fluid into blood and deep cervical lymph and its immunological significance. Brain Pathol 2(4):269–276

Cserr HF, DePasquale M, Harling-Berg CJ, Park JT, Knopf PM (1992b) Afferent and efferent arms of the humoral immune response to CSF-administered albumins in a rat model with normal blood–brain barrier permeability. J Neuroimmunol 41(2):195–202

Cutler RW, Page L, Galicich J, Watters GV (1968) Formation and absorption of cerebrospinal fluid in man. Brain 91(4):707–720

Davson H, Welch K, Segal MB (1987) Physiology and pathophysiology of the cerebrospinal fluid. Churchill Livingstone, Edinburgh

Dolan H, Crain B, Troncoso J, Resnick SM, Zonderman AB, Obrien RJ (2010) Atherosclerosis, dementia, and Alzheimer disease in the Baltimore Longitudinal Study of Aging cohort. Ann Neurol 68(2):231–240

Duong DH, O'Malley S, Sekhar LN, Wright DG (2000) Postoperative hydrocephalus in cranial base surgery. Skull Base Surg 10(4):197–200

Egnor M, Zheng L, Rosiello A, Gutman F, Davis R (2002) A model of pulsations in communicating hydrocephalus. Pediatr Neurosurg 36(6):281–303

Eller M, Williams DR (2009) Biological fluid biomarkers in neurodegenerative parkinsonism. Nat Rev Neurol 5(10):561–570

Engelhardt B (2008) The blood-central nervous system barriers actively control immune cell entry into the central nervous system. Curr Pharm Des 14(16):1555–1565

Engelhardt B, Ransohoff RM (2005) The ins and outs of T-lymphocyte trafficking to the CNS: anatomical sites and molecular mechanisms. Trends Immunol 26(9):485–495

Erlich SS, McComb JG, Hyman S, Weiss MH (1986) Ultrastructural morphology of the olfactory pathway for cerebrospinal fluid drainage in the rabbit. J Neurosurg 64(3):466–473

Fenstermacher JD (1984) Volume regulation of the central nervous system. In: Staub NC, Taylor AE (eds) Edema. Raven, New York, NY, pp 383–404

Foldi M (1975) Letter: lymphatic drainage of the brain. Lancet 2(7941):930

Foldi M (1996) The brain and the lymphatic system (I). Lymphology 29(1):1–9

Foldi M, Gellert A, Kozma M, Poberai M, Zoltan OT, Csanda E (1966) New contributions to the anatomical connections of the brain and the lymphatic system. Acta Anat (Basel) 64(4):498–505

Fox RJ, Walji AH, Mielke B, Petruk KC, Aronyk KE (1996) Anatomic details of intradural channels in the parasagittal dura: a possible pathway for flow of cerebrospinal fluid. Neurosurgery 39(1):84–90; discussion 90–91

Frankfort SV, Tulner LR, van Campen JP, Verbeek MM, Jansen RW, Beijnen JH (2008) Amyloid beta protein and tau in cerebrospinal fluid and plasma as biomarkers for dementia: a review of recent literature. Curr Clin Pharmacol 3(2):123–131

Galea I, Bechmann I, Perry VH (2007) What is immune privilege (not)? Trends Immunol 28(1):12–18

Geer CP, Grossman SA (1997) Interstitial fluid flow along white matter tracts: a potentially important mechanism for the dissemination of primary brain tumors. J Neurooncol 32(3):193–201

Glimcher SA, Holman DW, Lubow M, Grzybowski DM (2008) Ex vivo model of cerebrospinal fluid outflow across human arachnoid granulations. Invest Ophthalmol Vis Sci 49(11):4721–4728

Gomez DG, Chambers AA, Di Benedetto AT, Potts DG (1974) The spinal cerebrospinal fluid absorptive pathways. Neuroradiology 8(2):61–66

Gomez DG, DiBenedetto AT, Pavese AM, Firpo A, Hershan DB, Potts DG (1982) Development of arachnoid villi and granulations in man. Acta Anat (Basel) 111(3):247–258

Gomez DG, Fenstermacher JD, Manzo RP, Johnson D, Potts DG (1985) Cerebrospinal fluid absorption in the rabbit: olfactory pathways. Acta Otolaryngol 100(5–6):429–436

Greitz D (1993) Cerebrospinal fluid circulation and associated intracranial dynamics. A radiologic investigation using MR imaging and radionuclide cisternography. Acta Radiol Suppl 386:1–23

Grzybowski DM, Holman DW, Katz SE, Lubow M (2006) In vitro model of cerebrospinal fluid outflow through human arachnoid granulations. Invest Ophthalmol Vis Sci 47(8):3664–3672

Hammock MK, Milhorat TH (1976) The cerebrospinal fluid: current concepts of its formation. Ann Clin Lab Sci 6(1):22–26

Harling-Berg C, Knopf PM, Merriam J, Cserr HF (1989) Role of cervical lymph nodes in the systemic humoral immune response to human serum albumin microinfused into rat cerebrospinal fluid. J Neuroimmunol 25(2–3):185–193

Harling-Berg CJ, Knopf PM, Cserr HF (1991) Myelin basic protein infused into cerebrospinal fluid suppresses experimental autoimmune encephalomyelitis. J Neuroimmunol 35(1–3):45–51

Harnish PP, Samuel K (1988) Reduced cerebrospinal fluid production in the rat and rabbit by diatrizoate. Ventriculocisternal perfusion. Invest Radiol 23(7):534–536

Hatterer E, Davoust N, Didier-Bazes M et al (2006) How to drain without lymphatics? Dendritic cells migrate from the cerebrospinal fluid to the B-cell follicles of cervical lymph nodes. Blood 107(2):806–812

Hatterer E, Touret M, Belin MF, Honnorat J, Nataf S (2008) Cerebrospinal fluid dendritic cells infiltrate the brain parenchyma and target the cervical lymph nodes under neuroinflammatory conditions. PLoS One 3(10):e3321

Heisey SR, Held D, Pappenheimer JR (1962) Bulk flow and diffusion in the cerebrospinal fluid system of the goat. Am J Physiol 203:775–781

Hickey WF (1991) Migration of hematogenous cells through the blood–brain barrier and the initiation of CNS inflammation. Brain Pathol 1(2):97–105

His W (1865) Über ein perivasculäres Kanalsystem in den nervösen Central-Organen und über dessen Beziehungen zum Lymphsystem. Zeitschrift für Wissenschaft Zoologie 15:127–141

Johanson C (2008) Choroid plexus-CSF circulatory dynamics: impact on brain growth, metabolism and repair. In: Conn PM (ed) Neuroscience in medicine. Springer, Boston, pp 173–200

Johanson CE, Duncan JA III, Klinge PM, Brinker T, Stopa EG, Silverberg GD (2008) Multiplicity of cerebrospinal fluid functions: new challenges in health and disease. Cerebrospinal Fluid Res 5:10

Johnston M (2003) The importance of lymphatics in cerebrospinal fluid transport. Lymphat Res Biol 1(1):41–44; discussion 5

Johnston M, Papaiconomou C (2002) Cerebrospinal fluid transport: a lymphatic perspective. News Physiol Sci 17:227–230

Johnston M, Zakharov A, Papaiconomou C, Salmasi G, Armstrong D (2004) Evidence of connections between cerebrospinal fluid and nasal lymphatic vessels in humans, non-human primates and other mammalian species. Cerebrospinal Fluid Res 1(1):2

Johnston M, Zakharov A, Koh L, Armstrong D (2005) Subarachnoid injection of Microfil reveals connections between cerebrospinal fluid and nasal lymphatics in the non-human primate. Neuropathol Appl Neurobiol 31(6):632–640

Kida S, Pantazis A, Weller RO (1993) CSF drains directly from the subarachnoid space into nasal lymphatics in the rat. Anatomy, histology and immunological significance. Neuropathol Appl Neurobiol 19(6):480–488

Kida S, Okamoto Y, Higashi S et al (1994) Morphological aspects of interstitial fluid drainage from the rat brain. In: Nagai H, Kamiya K, Ishii K (eds) Intracranial pressure IX. Springer, Toyko, pp 136–139

Kida S, Weller RO, Zhang ET, Phillips MJ, Iannotti F (1995) Anatomical pathways for lymphatic drainage of the brain and their pathological significance. Neuropathol Appl Neurobiol 21(3):181–184

Kido DK, Gomez DG, Pavese AM Jr, Potts DG (1976) Human spinal arachnoid villi and granulations. Neuroradiology 11(5):221–228

Kleine TO, Benes L (2006) Immune surveillance of the human central nervous system (CNS): different migration pathways of immune cells through the blood–brain barrier and blood-cerebrospinal fluid barrier in healthy persons. Cytometry A 69(3):147–151

Kobayashi H, Minami S, Itoh S et al (2001) Aquaporin subtypes in rat cerebral microvessels. Neurosci Lett 297(3):163–166

Koh L, Zakharov A, Johnston M (2005) Integration of the subarachnoid space and lymphatics: is it time to embrace a new concept of cerebrospinal fluid absorption? Cerebrospinal Fluid Res 2:6

Koh L, Zakharov A, Nagra G, Armstrong D, Friendship R, Johnston M (2006) Development of cerebrospinal fluid absorption sites in the pig and rat: connections between the subarachnoid space and lymphatic vessels in the olfactory turbinates. Anat Embryol (Berl) 211(4):335–344

Koh L, Nagra G, Johnston M (2007) Properties of the lymphatic cerebrospinal fluid transport system in the rat: impact of elevated intracranial pressure. J Vasc Res 44(5):423–432

Leinonen V, Menon LG, Carroll RS et al (2011) Cerebrospinal fluid biomarkers in idiopathic normal pressure hydrocephalus. Int J Alzheimers Dis 2011:312526

Love JA, Leslie RA (1984) The effects of raised ICP on lymph flow in the cervical lymphatic trunks in cats. J Neurosurg 60(3):577–581

Lowhagen P, Johansson BB, Nordborg C (1994) The nasal route of cerebrospinal fluid drainage in man. A light-microscope study. Neuropathol Appl Neurobiol 20(6):543–550

Luetmer PH, Huston J, Friedman JA et al (2002) Measurement of cerebrospinal fluid flow at the cerebral aqueduct by use of phase-contrast magnetic resonance imaging: technique validation and utility in diagnosing idiopathic normal pressure hydrocephalus. Neurosurgery 50(3):534–543; discussion 43–44

MacAulay N, Zeuthen T (2010) Water transport between CNS compartments: contributions of aquaporins and cotransporters. Neuroscience 168(4):941–956

Manley GT, Fujimura M, Ma T et al (2000) Aquaporin-4 deletion in mice reduces brain edema after acute water intoxication and ischemic stroke. Nat Med 6(2):159–163

Mao X, Enno TL, Del Bigio MR (2006) Aquaporin 4 changes in rat brain with severe hydrocephalus. Eur J Neurosci 23(11):2929–2936

Marmarou A, Shulman K, LaMorgese J (1975) Compartmental analysis of compliance and outflow resistance of the cerebrospinal fluid system. J Neurosurg 43(5):523–534

McAllister JP II, Miller JM (2006) Aquaporin 4 and hydrocephalus. J Neurosurg 105(6 suppl):457–458; discussion 8

McComb JG, Hyman S (1990) Lymphatic drainage of cerebrospinal fluid in the primate. In: Johansson BB, Owman C, Widner H (eds) Pathophysiology of the blood brain barrier. Elsevier, Amsterdam, pp 421–438

McComb JG, Davson H, Hyman S, Weiss MH (1982) Cerebrospinal fluid drainage as influenced by ventricular pressure in the rabbit. J Neurosurg 56(6):790–797

McComb JG, Hyman S, Weiss MH (1984) Lymphatic drainage of cerebrospinal fluid in the cat. In: Shapiro K, Marmarou A, Portnoy H (eds) Hydrocephalus. Raven, New York, pp 83–98

Milhorat TH (1987) Cerebrospinal fluid and the Brain Edemas. Neuroscience Society of New York, New York, p 39–73

Moinuddin SM, Tada T (2000) Study of cerebrospinal fluid flow dynamics in TGF-beta 1 induced chronic hydrocephalic mice. Neurol Res 22(2):215–222

Mollanji R, Bozanovic-Sosic R, Silver I et al (2001a) Intracranial pressure accommodation is impaired by blocking pathways leading to extracranial lymphatics. Am J Physiol Regul Integr Comp Physiol 280(5):R1573–R1581

Mollanji R, Papaiconomou C, Boulton M, Midha R, Johnston M (2001b) Comparison of cerebrospinal fluid transport in fetal and adult sheep. Am J Physiol Regul Integr Comp Physiol 281(4):R1215–R1223

Mollanji R, Bozanovic-Sosic R, Zakharov A, Makarian L, Johnston MG (2002) Blocking cerebrospinal fluid absorption through the cribriform plate increases resting intracranial pressure. Am J Physiol Regul Integr Comp Physiol 282(6):R1593–R1599

Mollenhauer B, Trenkwalder C (2009) Neurochemical biomarkers in the differential diagnosis of movement disorders. Mov Disord 24(10):1411–1426

Nagelhus EA, Mathiisen TM, Ottersen OP (2004) Aquaporin-4 in the central nervous system: cellular and subcellular distribution and coexpression with KIR4.1. Neuroscience 129(4):905–913

Nagra G, Johnston MG (2007) Impact of ageing on lymphatic cerebrospinal fluid absorption in the rat. Neuropathol Appl Neurobiol 33(6):684–691

Nagra G, Koh L, Zakharov A, Armstrong D, Johnston M (2006) Quantification of cerebrospinal fluid transport across the cribriform plate into lymphatics in rats. Am J Physiol Regul Integr Comp Physiol 291(5):R1383–R1389

Nagra G, Li J, McAllister JP II, Miller J, Wagshul M, Johnston M (2008) Impaired lymphatic cerebrospinal fluid absorption in a rat model of kaolin-induced communicating hydrocephalus. Am J Physiol Regul Integr Comp Physiol 294(5):R1752–R1759

Nagra G, Wagshul ME, Rashid S, Li J, McAllister JP II, Johnston M (2010) Elevated CSF outflow resistance associated with impaired lymphatic CSF absorption in a rat model of kaolin-induced communicating hydrocephalus. Cerebrospinal Fluid Res 7(1):4

Naidich TP, Altman NR, Gonzalez-Arias SM (1993) Phase contrast cine magnetic resonance imaging: normal cerebrospinal fluid oscillation and applications to hydrocephalus. Neurosurg Clin N Am 4(4):677–705

Naruse I, Ueta E (2002) Hydrocephalus manifestation in the genetic polydactyly/arhinencephaly mouse (Pdn/Pdn). Congenit Anom (Kyoto) 42(1):27–31

Nitz WR, Bradley WG Jr, Watanabe AS et al (1992) Flow dynamics of cerebrospinal fluid: assessment with phase-contrast velocity MR imaging performed with retrospective cardiac gating. Radiology 183(2):395–405

Ohtani O, Ohtani Y, Li RX (2001) Phylogeny and ontogeny of the lymphatic stomata connecting the pleural and peritoneal cavities with the lymphatic system—a review. Ital J Anat Embryol 106(2 suppl 1):251–259

Oi S, Di Rocco C (2006) Proposal of "evolution theory in cerebrospinal fluid dynamics" and minor pathway hydrocephalus in developing immature brain. Childs Nerv Syst 22(7):662–669

Osaka K, Handa H, Matsumoto S, Yasuda M (1980) Development of the cerebrospinal fluid pathway in the normal and abnormal human embryos. Childs Brain 6(1):26–38

Owler BK, Pitham T, Wang D (2010) Aquaporins: relevance to cerebrospinal fluid physiology and therapeutic potential in hydrocephalus. Cerebrospinal Fluid Res 7:15

Papaiconomou C, Bozanovic-Sosic R, Zakharov A, Johnston M (2002) Does neonatal cerebrospinal fluid absorption occur via arachnoid projections or extracranial lymphatics? Am J Physiol Regul Integr Comp Physiol 283(4):R869–R876

Papaiconomou C, Zakharov A, Azizi N, Djenic J, Johnston M (2004) Reassessment of the pathways responsible for cerebrospinal fluid absorption in the neonate. Childs Nerv Syst 20(1):29–36

Phillips MJ, Needham M, Weller RO (1997) Role of cervical lymph nodes in autoimmune encephalomyelitis in the Lewis rat. J Pathol 182(4):457–464

Poca MA, Sahuquillo J, Busto M et al (2002) Agreement between CSF flow dynamics in MRI and ICP monitoring in the diagnosis of normal pressure hydrocephalus. Sensitivity and specificity of CSF dynamics to predict outcome. Acta Neurochir Suppl 81:7–10

Pollay M (2010) The function and structure of the cerebrospinal fluid outflow system. Cerebrospinal Fluid Res 7:9

Pollay M, Welch K (1962) The function and structure of canine arachnoid villi. J Surg Res 2:307–311

Preston JE (2001) Ageing choroid plexus-cerebrospinal fluid system. Microsc Res Tech 52(1):31–37

Preston SD, Steart PV, Wilkinson A, Nicoll JA, Weller RO (2003) Capillary and arterial cerebral amyloid angiopathy in Alzheimer's disease: defining the perivascular route for the elimination of amyloid beta from the human brain. Neuropathol Appl Neurobiol 29(2):106–117

Ransohoff RM, Kivisakk P, Kidd G (2003) Three or more routes for leukocyte migration into the central nervous system. Nat Rev Immunol 3(7):569–581

Rennels ML, Gregory TF, Blaumanis OR, Fujimoto K, Grady PA (1985) Evidence for a 'paravascular' fluid circulation in the mammalian central nervous system, provided by the rapid distribution of tracer protein throughout the brain from the subarachnoid space. Brain Res 326(1):47–63

Rennels ML, Blaumanis OR, Grady PA (1990) Rapid solute transport throughout the brain via paravascular fluid pathways. Adv Neurol 52:431–439

Romo-Gonzalez T, Chavarria A, Perez HJ (2012) Central nervous system: a modified immune surveillance circuit? Brain Behav Immun 26(6):823–829

Rubin RC, Henderson ES, Ommaya AK, Walker MD, Rall DP (1966) The production of cerebrospinal fluid in man and its modification by acetazolamide. J Neurosurg 25(4):430–436

Rudert M, Tillmann B (1993) Lymph and blood supply of the human intervertebral disc. Cadaver study of correlations to discitis. Acta Orthop Scand 64(1):37–40

Schley D, Carare-Nnadi R, Please CP, Perry VH, Weller RO (2006) Mechanisms to explain the reverse perivascular transport of solutes out of the brain. J Theor Biol 238(4):962–974

Seabrook TJ, Johnston M, Hay JB (1998) Cerebral spinal fluid lymphocytes are part of the normal recirculating lymphocyte pool. J Neuroimmunol 91(1–2):100–107

Shinohara H, Kominami R, Taniguchi Y, Yasutaka S (2003) The distribution and morphology of lymphatic vessels on the peritoneal surface of the adult human diaphragm, as revealed by an ink-absorption method. Okajimas Folia Anat Jpn 79(6):175–183

Silver I, Li B, Szalai J, Johnston M (1999) Relationship between intracranial pressure and cervical lymphatic pressure and flow rates in sheep. Am J Physiol 277(6 Pt 2):R1712–R1717

Silver I, Kim C, Mollanji R, Johnston M (2002) Cerebrospinal fluid outflow resistance in sheep: impact of blocking cerebrospinal fluid transport through the cribriform plate. Neuropathol Appl Neurobiol 28(1):67–74

Silverberg GD, Huhn S, Jaffe RA et al (2002) Downregulation of cerebrospinal fluid production in patients with chronic hydrocephalus. J Neurosurg 97(6):1271–1275

Spector R, Johanson CE (1989) The mammalian choroid plexus. Sci Am 261(5):68–74

Szentistvanyi I, Patlak CS, Ellis RA, Cserr HF (1984) Drainage of interstitial fluid from different regions of rat brain. Am J Physiol 246(6 Pt 2):F835–F844

Verkman AS, Binder DK, Bloch O, Auguste K, Papadopoulos MC (2006) Three distinct roles of aquaporin-4 in brain function revealed by knockout mice. Biochim Biophys Acta 1758(8):1085–1093

Vladic A, Klarica M, Bulat M (2009) Dynamics of distribution of 3H-inulin between the cerebrospinal fluid compartments. Brain Res 1248:127–135

Wagshul ME, Eide PK, Madsen JR (2011) The pulsating brain: a review of experimental and clinical studies of intracranial pulsatility. Fluids Barriers CNS 8(1):5

Weed LH (1914) Studies on cerebro-spinal fluid. No. II: the theories of drainage of cerebro-spinal fluid with an analysis of the methods of investigation. J Med Res 31(1):21–49

Wekerle H (1993) T-cell autoimmunity in the central nervous system. Intervirology 35(1–4):95–100

Welch K, Friedman V (1960) The cerebrospinal fluid valves. Brain 83:454–469

Welch K, Pollay M (1961) Perfusion of particles through arachnoid villi of the monkey. Am J Physiol 201:651–654

Welch K, Pollay M (1963) The spinal arachnoid villi of the monkeys Cercopithecus aethiops sabaeus and Macaca irus. Anat Rec 145:43–48

Weller RO (1998) Pathology of cerebrospinal fluid and interstitial fluid of the CNS: significance for Alzheimer disease, prion disorders and multiple sclerosis. J Neuropathol Exp Neurol 57(10):885–894

Weller RO, Kida S, Zhang ET (1992) Pathways of fluid drainage from the brain—morphological aspects and immunological significance in rat and man. Brain Pathol 2(4):277–284

Weller RO, Engelhardt B, Phillips MJ (1996) Lymphocyte targeting of the central nervous system: a review of afferent and efferent CNS-immune pathways. Brain Pathol 6(3):275–288

Weller RO, Subash M, Preston SD, Mazanti I, Carare RO (2008) Perivascular drainage of amyloid-beta peptides from the brain and its failure in cerebral amyloid angiopathy and Alzheimer's disease. Brain Pathol 18(2):253–266

Weller RO, Djuanda E, Yow HY, Carare RO (2009a) Lymphatic drainage of the brain and the pathophysiology of neurological disease. Acta Neuropathol 117(1):1–14

Weller RO, Preston SD, Subash M, Carare RO (2009b) Cerebral amyloid angiopathy in the aetiology and immunotherapy of Alzheimer disease. Alzheimers Res Ther 1(2):6

Weller RO, Galea I, Carare RO, Minagar A (2010) Pathophysiology of the lymphatic drainage of the central nervous system: implications for pathogenesis and therapy of multiple sclerosis. Pathophysiology 17(4):295–306

Widner H, Moller G, Johansson BB (1988) Immune response in deep cervical lymph nodes and spleen in the mouse after antigen deposition in different intracerebral sites. Scand J Immunol 28(5):563–571

Yamada S, DePasquale M, Patlak CS, Cserr HF (1991) Albumin outflow into deep cervical lymph from different regions of rabbit brain. Am J Physiol 261(4 Pt 2):H1197–H1204

Zakharov A, Papaiconomou C, Djenic J, Midha R, Johnston M (2003) Lymphatic cerebrospinal fluid absorption pathways in neonatal sheep revealed by subarachnoid injection of Microfil. Neuropathol Appl Neurobiol 29(6):563–573

Zakharov A, Papaiconomou C, Johnston M (2004a) Lymphatic vessels gain access to cerebrospinal fluid through unique association with olfactory nerves. Lymphat Res Biol 2(3):139–146

Zakharov A, Papaiconomou C, Koh L, Djenic J, Bozanovic-Sosic R, Johnston M (2004b) Integrating the roles of extracranial lymphatics and intracranial veins in cerebrospinal fluid absorption in sheep. Microvasc Res 67(1):96–104

Zhang ET, Richards HK, Kida S, Weller RO (1992) Directional and compartmentalised drainage of interstitial fluid and cerebrospinal fluid from the rat brain. Acta Neuropathol 83(3):233–239

Zlokovic BV, Segal MB, Davson H, Lipovac MN, Hyman S, McComb JG (1990) Circulating neuroactive peptides and the blood–brain and blood-cerebrospinal fluid barriers. Endocrinol Exp 24(1–2):9–17

Chapter 9
Collection of Human Limb Tissue Fluid and Lymph

W.L. Olszewski

Abstract The intercellular space mobile tissue fluid percolating cells contains all the information which proteins, carbohydrates, and lipids were filtered from plasma and what cells consume and excrete, as well as disintegrated cell debris. Moreover, it contains immune cells playing the immunosurveillance role. Harvesting of this fluid, that is called lymph, once it flows into lymphatics, gives insight into all tissue events. We worked out a microsurgical method of cannulation human leg lymphatics for continuous collection of lymph. This little invasive procedure allows collection of milliliters of lymph per hour. So far no local tissue reaction or edema of peripheral tissues was observed. Exclusion of one lymphatic collector from drainage of leg tissue doesn't affect lymphatic drainage due to many interconnections of peripheral lymphatics.

9.1 Introduction

Our understanding of tissue metabolic processes is based foremostly on investigations of blood percolating the given tissue. However, recent studies have shown that there are major differences in the biochemical composition of tissue fluid/lymph and plasma/serum. Tissue fluid contains products of plasma filtered through the capillary "molecular sieve." The low-molecular proteins are more easily transported across the capillary "membrane" than the large molecules (Olszewski et al. 1977). The effect is a different proportion of various proteins in tissue fluid/lymph compared with plasma. Moreover, a number of substances are

W.L. Olszewski (✉)
Department of Surgical Research and Transplantology, Polish Academy of Sciences, Warsaw 00950, Poland
e-mail: waldemar.l.olszewski@gmail.com

L. Santambrogio (ed.), *Immunology of the Lymphatic System*,
DOI 10.1007/978-1-4614-3235-7_9,

produced locally by the tissue parenchymatous cells. Mixed with those originating from plasma, their level will certainly be higher than in serum (Olszewski 2003). In addition, tissue fluid contains ground matrix compounds and subcellular structures from the senescent or damaged cells. All this makes necessary to get access to the intercellular space where the cell life can be observed in a natural context. This can be achieved by studies of the mobile tissue fluid the cells bathe in. The tissue fluid flows into lymphatics and changes its name to lymph. Continuous collection of lymph from cannulated lymphatics provides information at any time on the chemistry, cellularity, and hydraulics of fluid percolating tissue cells. Many studies carried out in man confirm what was observed in animal models or inferred from animal experiments. There are, however, also numerous original observations related specifically to human physiology or pathology that could be made only because the exploratory methods were extended to man. For example, observations of diurnal variations in protein concentration in leg lymph, protein concentration and output in leg lymph during physiological changes in capillary filtration, low and differentiated concentrations of various classes of immunoglobulins and complement proteins in lymph, retrograde transport of cholesterol from cells blood circulation via lymphatics, interstitial fluid pressure of skin of the limbs and subcutaneous tissue, and recirculation of lymphocytes through peripheral tissues in normal subjects and with chronic lymphatic leukemia, not to mention many others.

In man, lymph is more convenient to sample and study than interstitial fluid because the lymphatics are easily accessible and large volumes of lymph are available. There is, however, always a question of whether lymph is representative of the tissue fluid, and if so, to what extent. Studies have largely solved this problem, indicating that there are no significant differences between fluid protein levels in lymph and interstitial fluid, at least not in the skin and subcutaneous tissue.

For collection and study of human skin and subcutaneous tissue lymph and interstitial fluid, limbs seem to be the most suitable (Fig. 9.1). The material obtained originates from homologous tissue, and the influence of physical factors (capillary and venous pressure, external pressure, temperature), nervous and hormonal factors (nervous stimulation, hormones, drugs), and immune factors (infection, antigenic stimuli) on water and protein transport, as well as cell migration, can be studied. It should be emphasized that collection and in vivo studies of lymph and interstitial fluid must be carried out under uniform and strictly defined conditions. Many factors may considerably affect the volume and composition of lymph and interstitial fluid. These include the relationship of the location of the tissue from which the fluid is derived to heart level, metabolism, passive and active movements of the tissue, ambient temperature, temperature of the tissue, local and systemic nervous stimuli, drugs, water and electrolyte intake or output, trauma, and even the time of the day (circadian variations).

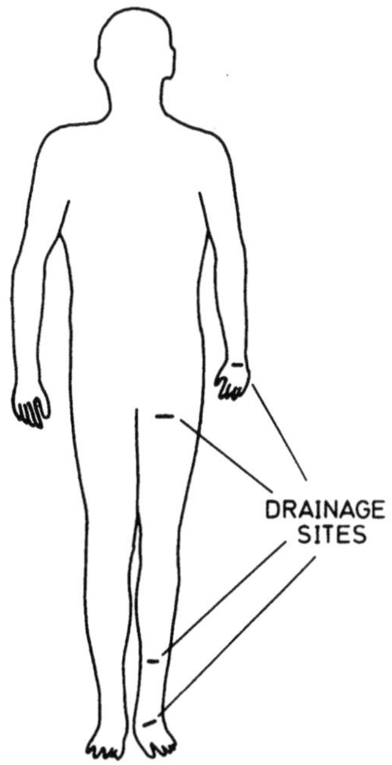

Fig. 9.1 Anatomical sites of location of superficial lymphatics most suitable for cannulation and injections, pressure measurements, and collections of lymph (adapted from (Olszewski 1991))

9.2 Anatomy of the Lymphatics of the Leg

The lymphatic trunks of the leg are divided into a deep and a superficial system. The latter consists of the medial and lateral groups. In the lower leg, the medial superficial group is composed of two to three collecting trunks. They cross the long saphenous vein at the level of the knee but are never close to it and terminate in the inguinal lymph nodes. The lateral superficial group consists of one to two trunks that are apposed to the short saphenous vein and run toward the popliteal fossa. The superficial system drains lymph from the digital lymphatic plexuses, foot and calf skin, and subcutaneous tissue. In the calf, the deep system is formed by the union of the posterior tibial and fibular trunks which anastomose with a separate anterior tibial trunk. It collects lymph from the periosteum, fasciae, and muscles.

There is a strictly regional pattern of drainage. Injection of dye into the web space of the hallux shows the medial superficial group that runs into the subcutis of the lateral aspect of the heel and the lateral superficial trunk. The deep lymphatics are visualized by injecting dye into the deep structures of the sole of the foot. In the normal individual, no anastomoses (Fig. 9.1) can be demonstrated between the subfascial and epifascial lymphatic trunks. Thus, lymph sampled from a collecting trunk in the calf is derived from a defined area and mass of tissues. The medial superficial group trunks are most suitable for chronic cannulation in the lower limb (Fig. 9.2a).

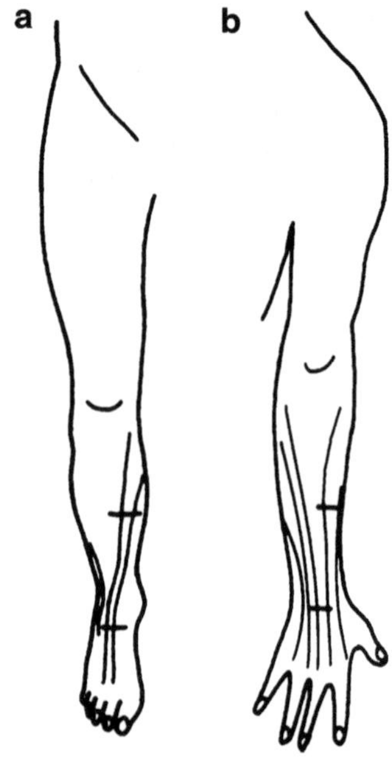

Fig. 9.2 Anatomy of limb lymphatics. (**a**) Leg; (**b**) forearm (adapted from (Olszewski 1991))

9.3 Anatomy of the Lymphatics of the Upper Limb

The digital lymphatic plexuses are drained by vessels that run along the outer margin of the fingers toward the web, where they are joined by vessels from the distal palm and pass toward the dorsal surface of the hand. Collecting vessels from the palmar plexus unitę to form a trunk that runs around the metacarpal bone of the fifth finger to join the dorsal vessels of the same finger. The ulnar group of lymphatics drains the third, fourth, and fifth fingers as well as the ulnar side of the hand and forearm. These vessels run along the basilic vein to the cubital region. The radial group of lymphatics drains the first and second digits and the radial side of the hand and forearm. Lymphatics running on the dorsal radial part of the forearm are most convenient for chronic cannulation (Fig. 9.2b).

9.4 Chronic Cannulation of the Lymphatics of the Leg

A method for chronic cannulation of leg lymphatics has been worked out by Engeset et al. (1973). Using this method, Olszewski (1977) cannulated thigh and foot lymphatics in patients with pathological alterations of the peripheral lymph vessels. Later this technique was used for cannulation of normal leg lymphatics in studies on the physiology of lymph flow and its biochemistry, as well as penetration of antibiotics and chemotherapeutics to tissue fluid and reversed cholesterol transport from tissues to blood circulation (Olszewski et al. 2001; Bergan et al. 1982; Nanjee et al. 2000). Leg lymph can be easily obtained by cannulation of the superficial collecting trunks on the dorsum of the foot, in the medial group of calf vessels, and in the thigh and groin. For chronic cannulation, however, the lower part of the calf seems to be most suitable.

After injection of patent blue violet (PBV) in the same way as was used time ago for routine lymphography, the lymph vessel is exposed. However, PBV coloration of the lymph may influence the results of biochemical studies when colorimetric methods are used; lymph cells may also be damaged. It is preferable to use no dye if lymph composition studies are planned. With experience, lymphatic trunks can be found easily, even without coloration.

The cannulation is carried out under strictly sterile conditions. Skin is anesthetized with 1 % Xylocaine, and a 2 cm incision is made at the front aspect of the calf about 10–15 cm above the ankle. Further steps are made with the use of the operating microscope under magnification 4–20 times. The vessel is exposed and the sterile, siliconized polyethylene Clay-Adams P60 tube with a conical tip filled with heparin saline solution is advanced distally through the opening made with the use of microscissors. Contamination of the tube tip with wound content should be avoided in order to prevent formation of clot in the tube after it is introduced into the lymphatic vessel. A fine thread is put around the vessel containing the tube and tied gently so that lumen is not occluded. The tube is also fastened to the surrounding tissues in the wound with one to two sutures (Fig. 9.3). The wound is closed, and the tube is fixed to the skin with adhesive tape. The external tip of the tube is inserted into the sterile disposable syringe through its needle outlet. The plunger is displaced distally so that the whole syringe can be gradually filled with lymph (Fig. 9.4).

The use of syringes has an advantage over other methods because it protects the sterility of lymph and allows an accurate measurement of volume in the course of collection. The danger of lymph evaporation is also avoided since the tube almost occludes the needle outlet of the syringe. As required, heparin solution, culture media, and others can be placed at the bottom of the syringe. Because the syringe is fixed to the leg with adhesive tape, patients are allowed to walk, and there is practically no danger that the tube will slip out. Collection of lymph can be carried out continuously for several days. Special care should be exercised to avoid contamination of the wound and tube during syringe changes. There may be some clotting of lymph in the tube, especially in cases of Iow lymph flow. The clot usually protrudes from the tube and can easily be removed by gentle traction with forceps. The temperature of lymph collected in syringes fixed to the calf skin is around 27–31 °C. If a Iow temperature is required, the collecting syringes can be placed in isopore boxes filled with ice and fixed to the leg. After completion of lymph collection, the cannula is removed by pulling and a pressure dressing is put at the incision site for 24 h. Cannulation of leg lymphatics for periods as long as several weeks is not burdened by complications. With experience of over 1,500 cannulations, the author did not observe wound infections, lymph leakage, and postcannulation lymphedema. The cannulated vessels become obliterated and the neighboring vessels are taken over the drainage. Cannulation of the upper limb lymphatics is practically the same as that described for the lower limb.

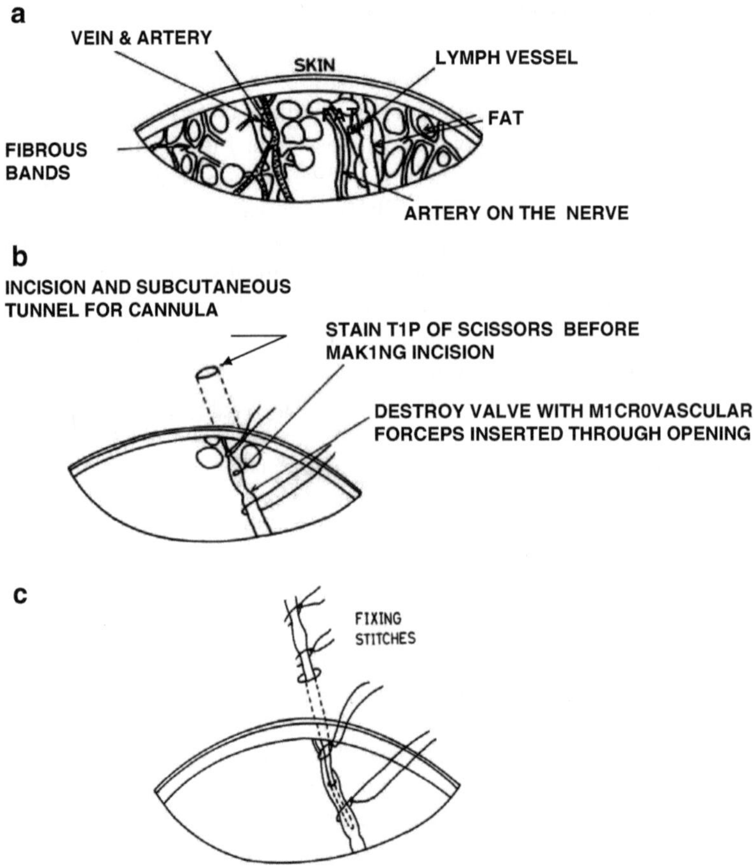

Fig. 9.3 Diagram of the technique of cannulation of lymphatics in human leg for lymph collection, angiography, drug infusion, and pressure and flow measurement. (**a**, **b**, and **c**) Stages of the operation (for details see text) (adapted from (Olszewski 1991))

Fig. 9.4 Method of collection of lymph in calibrated syringes (adapted from (Olszewski 1991))

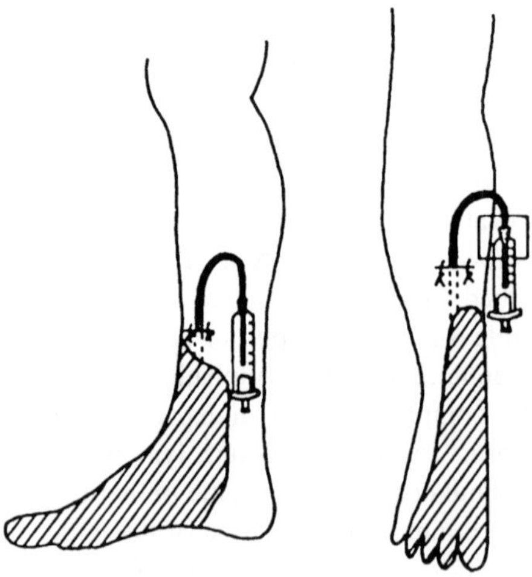

9.5 Other Methods of Limb Lymph Collection

Lymph can also be collected by direct puncture of intact cutaneous lymphatics. Briefly, PBV is injected intradermally with a 30-gauge needle into the region where lymph is to be collected. After 10–20 min, blue streamers are produced. A 23- to 30-gauge needle connected to plastic tubing is then introduced into the dermal lymphatic counter to the flow current. Use of magnifying glasses facilitates the procedure. The syringe is used for gentle aspiration. However, care must be taken not exert a pressure that would collapse the vessel and draw air into the tubing. The fluid is collected in the tubing and its volume can be measured directly. With this method, only several microliters of lymph can be collected.

In cases of lymphedema with dilated cutaneous and subcutaneous lymphatics, especially the hyperplastic type, lymph samples can be obtained by direct puncture of the skin. Several 23-gauge needles with splints connected to plastic tubing are inserted into the skin and left for several minutes to allow bleeding to stop. The splints are gently removed, and the needles are withdrawn 1–2 mm and are left for 1–2 h. With this method, several microliters of interstitial fluid and lymph can be obtained from each needle, even without aspiration. Contamination of lymph with blood should be avoided. The hemoglobin concentration should be measured in each sample.

9.6 Collection of Lymph for Biochemical and Cellular Studies

Generally, it is possible to collect lymph in sufficient amounts for protein, electrolyte, acid–base balance, and gasometric studies. The concentrations of immunoglobulins, complement, lymphokines, and other immune proteins can also be measured with routine techniques. Since the collection time for lymph is usually several hours and the temperature in the collecting syringes is around 30 °C, the possible inactivation of enzymes and other temperature-unstable substances should be taken into account. For collection of lymph cells, heparin without preservatives should be added to the collecting syringes in concentrations not exceeding 5 U/mL. To avoid adherence of dendritic cells to the walls of the cannula and syringe, the wall surfaces should be siliconized (Figs. 9.3 and 9.4).

9.7 Remarks on Lymph Flow and Content

Lymph flow is measured by determining the volume of lymph collected in the calibrated syringe during one unit of time or using a specially designed lymph flow meter. Since accurate estimation of the weight of tissues from which lymph is derived is difficult, flow rate cannot be presented in units of volume time tissue weight. This makes it difficult to compare interindividual flow data. There are significant differences in lymph flow between individuals, and from leg to leg in the same individual (due to the varying topography), caliber of the cannulated vessels, and permeability of blood exchange vessels.

As mentioned in introduction, the lymph concentration of proteins, lipids, and carbohydrates significantly differs from that of plasma. Generally, it is a plasma filtrate inversely proportional to the molecular weight and radius of filtered substances. Moreover, it changes within minutes depending of the capillary filtration rate that in turn is dependent on the intracapillary hydrostatic pressure, local metabolic processes, temperature, and the compliance of the interstitial space. The 1-h collection samples give different results from the 2-h sample. The 24 h collection samples are recommended for biochemical and cellular studies. Food and fluid intake should be standardized. The administration of medicines should be controlled as they may affect blood capillary flow and cause release of regulatory peptides from neurohormonal tissues fast penetrating to the tissue fluid.

References

Bergan T, Engeset A, Olszewski W, Josefsson K, Larsen N (1982) Penetration of erythromycin into human peripheral lymph. J Antimicrob Chemother 10:319–324

Engeset A, Hager B, Nesheim A, Kolbenstvedt A (1973) Studies on human peripheral lymph. I. Sampling method. Lymphology 6:1–5

Nanjee MN, Cooke CJ, Olszewski WL, Miller NE (2000) Lipid and apolipoprotein concentrations in prenodal leg lymph of fasted humans. Associations with plasma concentrations in normal subjects, lipoprotein lipase deficiency, and LCAT deficiency. J Lipid Res 41:1317–1327

Olszewski WL (1977) Collection and physiological measurements of peripheral lymph and interstitial fluid in man. Lymphology 10:137–145

Olszewski WL (1991) Collection and physiological measurements of peripheral lymph and interstitial fluid in man. In: Olszewski WL (ed) Lymph stasis: pathophysiology, diagnosis and treatment. CRC Press, Boca Raton, pp 609–615

Olszewski WL (2003) Pathophysiological aspects of lymphedema of human limbs: I. Lymph protein composition. Lymphat Res Biol 1(3):235–243

Olszewski WL, Engeset A, Sokolowski J (1977) Lymph flow and protein in the normal male leg during lying, getting up, and walking. Lymphology 10(3):178–183

Olszewski WL, Pazdur J, Kubasiewicz E, Zaleska M, Cooke CJ, Miller NE (2001) Lymph draining from foot joints in rheumatoid arthritis provides insight into local cytokine and chemokine production and transport to lymph nodes. Arthritis Rheum 44:541–549

Index

L. Santambrogio (ed.), *Immunology of the Lymphatic System*,
DOI 10.1007/978-1-4614-3235-7, © Springer Science+Business Media New York 2013

MIX
Papier aus verantwortungsvollen Quellen
Paper from responsible sources
FSC® C105338

If you have any concerns about our products,
you can contact us on
ProductSafety@springernature.com

In case Publisher is established outside the EU,
the EU authorized representative is:
Springer Nature Customer Service Center GmbH
Europaplatz 3, 69115 Heidelberg, Germany

Printed by Libri Plureos GmbH
in Hamburg, Germany